The
LAMBTON WORM

The
LAMBTON WORM
The Definitive Guide to Angling in North East England

PETE MCPARLIN

AMBERLEY

First published 2011

Amberley Publishing
The Hill, Stroud
Gloucestershire, GL5 4EP

www.amberleybooks.com

British Library Cataloguing in Publication Data.
A catalogue record for this book is available from the British Library.

ISBN 978-1-4456-0454-1

Typesetting and Origination by Amberley Publishing.
Printed in Great Britain.

Contents

PART TWO: NORTH EAST FISHING SEASONS

Introduction

To outsiders, it would seem that the North East of England is the forgotten area of the British Isles when it comes to angling. Talk about game fishing and thoughts turn immediately to southern chalkstreams and Scottish rivers – of which the Tweed is, at least, partly North Eastern – while famous Northumbrian streams like the Tyne, Wear and Coquet are all too often overlooked.

Coarse fishing will conjure up pictures of virtually every part of the country other than the North East – ponds and lakes in lowland southern England that contain giant carp and tench, and rivers like the Hampshire Avon and Severn, the stuff of which legends are made.

Even the North East's sea anglers, whose traditions are possibly strongest in the northernmost counties of England, are often overshadowed by those from places like the south-west and the west coast of Scotland, not to mention almost the whole of Ireland!

Yet to we who live in this area, the rivers, lakes, ponds, beaches and rocky coastline of our home region offer as varied and high-quality fishing as that which is found anywhere else on these islands. And those who have visited – and fished – know that the North East, with its many reservoirs, is a great place to come for both the experienced and the casual angler alike.

For in truth, we are ahead of the game in this ancient kingdom of Northumbria. The Tyne is now generally accepted to be the finest river in England for salmon fishing and many would argue that the Tweed fulfils a similar brief for the UK as a whole – if not the world! The Wear is fast emulating the Tyne for the quality of its sea trout angling, while the smaller rivers, Coquet and (Yorkshire) Esk, 90 miles apart, have well-established reputations for their runs of both migratory species.

The brown trout is literally ubiquitous in the waters of North East England. There are very few rivers or lakes where it cannot be found; even the tiniest burn can sometimes harbour a monster lurking under the piles of a road bridge – such is the virility of the insect life in the area's rivers and streams. Just take an early evening walk within a few hundred yards of anything more substantial than a drainage ditch in early June and see how many mayflies you can spot dancing in the shafts of sunlight amid the tree branches. Our native trout can vary in size from anything from a few ounces, in the smaller rivers and streams, up to several pounds in the larger watercourses. Some of them will even migrate to sea and return as fish of up to twenty pounds to the rivers Tyne, Wear and Coquet.

Then there are those reservoir trout, which can sometimes be caught at weights approaching that of their migratory cousins. The hilly nature of Northumbria's inland areas has resulted in no fewer than eighteen reservoirs that have been built to supply water in Northumberland and Durham (including the vast Kielder), with a further seven on the North York Moors and in the northern Yorkshire Dales. The water companies that own them have not been slow to realise their fishing potential and anglers now flock from far and wide to fish for rainbow and brook trout at places like Fontburn and Derwent Reservoirs.

Further south, the coarse fishing offered by the rivers Swale, Ure and other tributaries of the Yorkshire Ouse has long been regarded as the some of the finest in the land. The majestic barbel is rightly admired for its fighting prowess by those lucky enough to have tangled with one and, unlike more famous fisheries, such as the Severn or the Avons of Hampshire, Warwickshire or Bristol, Yorkshire's barbel are indigenous, having migrated here themselves following the last ice age. Ditto the chub, roach, dace, perch, pike and grayling – all natural to the flowing waters of the Swale and Ure, as well as the Tees, Wear and, to a lesser extent, the Tyne.

For the varied nature of our region's fishing, we need look no further than the geographical relief of the area, which consists, from west to east, of hill ranges (the Cheviots in the north, Pennines in the south), the foothills thereof and a coastal plain; this is uniform almost all the way from north to south. The result is more or less every type of spate river and lake from the greatest – the rivers Tyne, Tees and Tweed and Kielder Reservoir – to the smallest – the Aln, Blyth and Esk, Whittle Dene Reservoir and the numerous ponds and other small stillwaters dotted throughout the area. The undulating coastline gives rise to great cliffs, such as those at Dunstanburgh, in Northumberland, and numerous storm beaches from which winter cod are still caught in good numbers. For the all-round angler, or even the specialist, this must seem like God's own angling country, but it is only available to those prepared to come and explore.

Back in the mid-1970s, the noted North East angling journalist Charles Wade wrote:

Before the last Ice Age, Northumbria's three main rivers, the Tyne, Wear and Tees, were one massive watercourse and, with the disturbed land and shifting boulders, there eventually settled down a pattern of rivers which are second to none.

The South Tyne at Haydon Bridge.

While this observation isn't strictly true to what we now think probably happened, Charles was certainly ahead of the accepted wisdom of his time. Nowadays, many geologists believe that, prior to the ice age, the River Wear was indeed a tributary of the main Tyne, while the Tees may well have been connected to at least some of the main Yorkshire rivers, namely the Swale and the Ure.

Indeed, several pieces of evidence point to a compelling argument for that unholy alliance between Wear and Tyne, despite what many purists from both localities may wish to believe. Take a look north from the small hilltop village of Edmondsley, just 5 miles from Durham City, and the Wear valley can be seen quite obviously continuing on its northward bearing until it meets the Tyne – the tower blocks of western Newcastle are clearly visible on the north bank of the main river from well over 10 miles away.

The Wear itself doesn't take this course, needless to say – it veers eastwards at Chester-le-Street to reach the sea at Sunderland. The wide, deep valley continuing north to meet the Tyne facing Elswick is instead occupied by the diminutive River Team, a trickle that gives its name to the sprawling valley bottom trading estate near Gateshead – a river that runs little more than ten miles from source to confluence and one that can be jumped over for all but the last three to four hundred yards of its course. Surely this couldn't be the river that gouged out the huge depression that you drop into after passing The Angel on the A1 going north?

Indeed it isn't. Geologists now think that the terminal moraine of the Wear valley glacier blocked the course of its melting waters, some fourteen thousand years ago, diverting the re-emergent River Wear through a gap in the east Durham magnesium limestone ridge at what is now Monkwearmouth. Thus, the Wear gained a new course, and one which would have obvious, if unforeseen, disadvantages in the millennia to come. The un-glaciated estuary of the lower Wear was at a distinct disadvantage to its fierce rival, come the height of the shipbuilding industry in the late nineteenth century. While the aggregate tonnage of ships launched into each estuary was similar, the 'new' river's shallow bedrock limited the depth to which it could be dredged and, hence, the size of the steel-hulled ships that could be launched.

Such points are not of issue to the angler, of course, but other geological factors added other post-industrial problems – matters that are often still cause for concern to this day. 'River capture', a process in which the upper reaches of river valleys were appropriated by other watercourses, happened long before the last ice age, and resulted in the Wansbeck and the Blyth, to name but two, flowing through far deeper valleys than these modest-sized rivers ought to have made. The 'thieves' in these instances are the North Tyne and Rede (take a look on the map and you'll see!), rivers whose more rapid head-ward erosion eventually carved back the Cheviot Hills until they took their neighbours' upper valleys for their own.

While this geographical equivalent of border rieving was taking place, the first coal measures were being laid down beneath the upper rock strata, establishing a raw material upon which the North East's industrial heritage would eventually depend. Coal mining began in the area before Roman times and by the mid-twentieth century, between 20 and 30 million tons of coal a year was being extracted from the Durham coalfield alone.

The nature of deep mining, which commenced in the early nineteenth century, necessitated pumping to clear the workings of groundwater. This water, usually rendered poisonous by heavy metal deposits, could amount to thousands of gallons per minute and was simply discharged into the nearest river or stream, with horrendous consequences to wildlife, including fish. The River Wear was probably the worst affected of the Northumbrian rivers, a situation compounded from the 1830s when mines in east Durham started being sunk through water-laden Permian sand measures.

When the last pits finally closed towards the end of the twentieth century, the problem still didn't go away. To prevent mine water from flooding the abandoned workings and bringing far greater concentrations of toxic compounds to the surface, the pumps had to be kept switched on throughout the former Durham coalfield in order to protect the fragile ecosystem of the rejuvenated River Wear. In 1994, it was estimated that 25 million gallons of water per day were being pumped, at great cost to the coal authorities, who were keen, it was known, to switch off the pumps and save themselves tens of millions of pounds. Pressure from anglers and environmental groups helped prevent an ecological catastrophe, but the threat will remain for decades to come.

Lead mining in the high North Pennine dales can be traced back to the middle ages and has been equally damaging to certain northern rivers as its lowland compatriot. The lead mining industry peaked in the mid to late nineteenth century and the principal victims of this polluter were the South Tyne and that perennial casualty, the Wear.

As with coal mining, pollution was principally caused by groundwater, although, with most of the workings being relatively shallow, the problem wasn't related to pumping water out. With lead mines, it was more a case of water going in – feeder streams were diverted to run through the workings, helping the miners in their arduous task of extracting ore from the rock. However, this water still had to go somewhere and inevitably it made its way into the main rivers, now laden with lead and other toxins that poisoned all aquatic life.

Lead mining finally ended in the area at the start of the twentieth century and with its decline there was a marked improvement in water quality in the upper reaches of the rivers affected. Other pollution problems also came and went, most notably the spoilage of the Wear, Tees and, in particular, the Tyne estuaries by raw sewage discharge. These problems have also now been overcome and, along with reductions in coastal discharges, fresh and salt water across the region is almost all now of a quality sufficient to support all forms of life.

And so, the story of angling in the North East has come full circle. The rivers, lakes and sea are all now as clean as they have been for the best part of 250 years, and the invertebrate and plant life, as well as the fish and water fowl that feed off them, are back to full health. The range of species that can be fished for in the area is now even greater, thanks to the introduction of species like the rainbow trout, and even the effects of global warming have contributed – bass are now being caught from the beaches of Northumberland and Durham for the first time in recorded history. Of course for those who already fish here, this is hardly news, but for anyone who hasn't, why not come and have a go? Like me, you might just find it is the greatest fishing experience of your life!

PART ONE

The Essence of North East Fishing
– Past & Present

CHAPTER ONE

A River Runs Through It: The History of River Trout Fishing in the North East

A century and a half ago, as angling became as much a sport as an act of necessity, wealthier folk began to regard fishing for that spotted freshwater fish, the trout – more palatable and often found in more exclusive locations – as somehow more righteous than angling for their roughly scaled brethren, species such as the dace and the chub. The demarcation that still exists to this day was thus established, with the trout to become known as a game fish: a superior inhabitant of streams, fished for by a 'superior' class of angler!

In what was an otherwise enlightening book, *Angling; Or, How to Angle and Where to Go*, the nineteenth-century angling writer Robert Blakey epitomised this attitude. 'Next in importance to the salmon, in the estimation of the genuine angler, stands the trout,' he wrote. 'He is the standard commodity of the enthusiastic rod fisher. There are many expert and experienced fly fishers who never enjoyed the unique and exciting luxury of hooking and killing a salmon; but no man can fairly lay claim to the appellation of an "angler", if he cannot kill trout with the rod and line in some way or another.'

Launching into a characteristically Victorian salvo of anthropomorphism, Blakey continued: 'There is something about trout fishing which has exalted it in all eyes above every other branch of the art, except, of course, that of salmon fishing. If we attempt to analyse this preference, we shall find it resolve itself into something appertaining to the attributes, qualities, or habits of this beautiful and interesting fish. He is an intellectual kind of creature, and evidently has a will of his own – he looks sagacious and intelligent; he sedulously avoids thick, troubled and muddied waters, displays an ardent ambition to explore the rivers to their very source; is quick, vigorous and elegant in his movements – is comparatively free from vulgar, low and grovelling habits; and, in a word, in every stage of his existence, preserves a superior and dignified demeanour unattainable by any other occupant of the streams. These may be styled the social and intellectual qualities of this glorious fish.'

Nowhere more so did this angling doctrine take hold than in Blakey's native North East of England, and while the dogma of the upstream dry fly – so prevalent on southern chalk streams – may have been impractical on most northern rivers, in many places angling for game fish nonetheless became a rigorously elite practice. For the next hundred years from the early 1800s onwards, this ideal would govern fishing in Northumbria, where it became most apparent with the discipline of game angling.

Thankfully, such elitism did die out, and the increase in wealth among the working and middle classes eventually led to the formation of angling clubs and associations. In most cases, these organisations strove to provide affordable fishing for all and, as landowners began to lease the fishing rights to these 'preservers of the water', by the mid-twentieth century much of Northumbria's prized game fishing was within the reach of the ordinary person. Trout fishing had gone full circle and was once again the preserve of the common man.

So why exactly is it that game fishing is so pre-eminent in the North East region? Well firstly, unlike almost anywhere else in England, the native British trout, *Salmo trutta*, is the dominant species from source to sea in every river in Northumberland and Durham. While the largest rivers in the area, the Wear and the Tyne, do contain naturally occurring coarse species in their lower reaches, unlike in rivers further south, they never quite manage to take over. Cast a worm into the mid to lower reaches of either of these Northumbrian rivers and there is greater than even chance that it will be taken by a trout. Do the same thing on the lower Swale or Ure – rivers not all that much further south – and the likelihood is in catching something entirely different: on these rivers coarse fish begin to take over the further downstream from their dales that you go.

On the smaller rivers of Northumberland, the brown trout is omnipotent – there is little likelihood of catching anything else in the Wansbeck or the Blyth, and absolutely no chance of other species on the Coquet or Aln – except for salmon and eels. And while the majestic River Tweed does contain some less regal species, these are believed to have been introduced by man at some point in the past. At times, the dace, roach and grayling have all flourished in the lower Tweed, the Teviot and the Till, but the trout have always been by far greater in number.

To further explain the brown trout's dominant presence in this area, we need look no further than the geography of the region and the way that this selects biologically in favour of the trout and its lifestyle. All the streams of Northumbria are spate rivers, all rising in upland areas, with the largest starting life high on mountainous moorland. The water in the headwaters is always cold, peaty and acidic and, flowing out of crevices in the limestone bedrock, it is largely devoid of nutrients. With few exceptions (none of which are large enough to be considered angling species) the brown trout is the only fish able to survive in such conditions, with the sparse invertebrate life placing a stringent nutritional control on the size to which any fish can grow.

A common characteristic of such rivers, particularly the largest ones, is a good run of sea trout. These are simply brown trout which migrate to sea, usually from the sparse upland feeder streams where food is so scarce, to put on weight well in excess of what they could achieve if they stayed in the river, and then return to their natal headwaters to spawn. A sea trout can do this any number of times throughout its life and the offspring can be either migratory or not. Hence, in addition to trout spawned by those brown trout that don't migrate, any spate river will normally maintain a healthy population of the species, proportional to its size.

Another factor is climate. The northerly latitude of the North East region, combined with the cooling effect of winds off the North Sea, means that coarse species such as dace, chub and roach are near the natural limit of their temperature tolerance, even in the lowland waters of the Wear and Tyne. Unlike the rivers Tees, Swale and Ure, which descend quickly from the Pennines before slowing considerably, the two largest Northumbrian rivers are relatively swift throughout their courses, except for the tidal reaches. This tends to further oxygenate the lower reaches and also selects biologically towards the habitat preferences of the trout over other species.

Needless to say, trout fishing with rod and line has been of importance for countless generations in the North East – originally as a valuable food source for communities located close to rivers, and not always by means considered 'fair' by those nineteenth-century grandees. A painting dating back to the 1820s by the artist T. M. Richardson, showing Barras Bridge – then located just outside Newcastle – clearly depicts two young boys fishing in one of the small streams that ran down to the Tyne close to the city walls. Given the gradient at which what is now a storm drain descends to the main river, the drop must have meant that the only species able to negotiate such precipitous flows was the trout.

Robert Blakey wrote of another such young lad's trout, caught on the Coquet in the 1820s beneath the crossing-point of the main road into Scotland: 'Nearly about the same time (the dry summer of 1826), a large trout, under precisely the same circumstances, was observed for a long time near to one of the arches of Felton Bridge. He took up a sort of permanent abode there; had often anglers paying him a visit, but all their subtle arts proved unavailing, and he was captured at last by a country lad, with a miserable rod and line, with a plain red worm. His weight was five pounds.'

A little over a century later, another young angler was frequenting the nearby River Wansbeck, also in pursuit of brown trout and also with the most rudimentary of tackle. In his autobiography, the 1966 World Cup hero Jack Charlton nostalgically recalled his own boyhood fishing exploits, often with younger brother Bobby in tow:

When I was perhaps twelve or thirteen, I discovered that there were trout
in the river at Bothal, about a couple of miles upstream from Ashington.
I'd been to the place before for minnows, but one day I saw this huge
trout. We weren't allowed to fish for them, of course, but I said to myself,
hell, I'm going to have one of those before long.

The more practically minded of the Charlton brothers quickly hatched a plan:
'I soon figured out how to do it. I'd take a ball of string, attach a piece of
wood, and then tie on a dropper with a worm on the end. Then I'd retreat
round the corner, so nobody could see me and wait. I'd sometimes catch as
many as ten trout a night. It was all totally illegal, but I was growing up fast in
a hard world, and I never got caught.'

Jack's recollections must strike a chord with at least one long-standing
member of the Wansbeck Angling Association, who remembered his own
youthful perambulations on those same banks, many years later. 'As a
Pegswood lad, I fished the more prolific waters at Bothal and the Viaduct
Woods,' he remembered, revealing that, 'The catches were generally much
weightier than those taken upstream at Morpeth.'

He thus betrayed a degree of local knowledge vital for a better share of the
biggest trout in the river – information sadly lost on at least one of his Wansbeck
contemporaries. In fact, as far back as 1859, Robert Blakey had warned, 'In the
rivers Rede, Blyth and Wansbeck, there are fine trout, but they can only be
properly angled for by persons who have a very accurate knowledge of the
peculiarities of each stream. For general tourists they are not best suited.'

Indeed, in the formative years of Wansbeck Angling Association, only
residents of the towns and villages adjacent to the river were permitted
to become members, a rule common to many North Eastern clubs and
associations at the time. Another club that for many years only permitted
locals to join was the Derwent Angling Association, whose membership was
restricted to residents of Consett, Shotley Bridge and pit villages further down
the valley, such as Chopwell. At the DAA's peak in the 1960s and 1970s, there
was a waiting list of upwards of two years and such was the demand that,
upon acceptance, a new member was required to report to the local post office
with ID in order to collect his or her permit.

Other rules included restrictions on the baits and lures that could be used
during the trout fishing season, regulations that for the most part are still in
place today. On most rivers in the Northumbria region – and indeed on all
association waters on the Wansbeck – there was a stipulation that, before
1 June, anglers were permitted to fish for trout by 'fly only', meaning (as on
those rivers that did contain coarse fish the season was closed from mid-March
to mid-June) that no other bait or lure was allowed for the first two and a half
months of the trout season.

An early season brown trout. The fish's size, as indicated by the rule, shows that it is takeable.

Fly-only rules were sometimes extended throughout the season by certain clubs, notably the Derwent Angling Association, while at others no live bait other than worm or, sometimes, minnow was allowed after 1 June. From 1974 onwards, many of these regulations became underpinned by byelaws, under the Salmon and Freshwater Fisheries Act, but there were loopholes and there were also those despicable individuals who paid no heed to the rulebook in any case.

For those prepared to go by the book back in the mid-1970s, Frank Johnson's *North East Angling Guide* was probably the best example, and it included an inventory of the standard North Eastern trout angler's fly box, as recommended by J. L. Hardy of the famous Alnwick tackle firm, Hardy Bros. Hardy suggested, 'It is useful to know which type of fly will generally be taken under certain conditions and the following list of flies will be most useful on any north country trout stream.'

Snipe and Purple: This is a splendid killer on cold days in the early part of the season, particularly in March and April when there are often hail showers and cold winds.

Waterhen Bloa: A useful fly in March and April and at the end of the season, particularly on cold windy days, when they are found to linger on the surface. Also a very useful grayling fly.

Dark Needle or Needle Brown: Good killer all of the season, particularly on days of flying clouds and fitful bursts of sunshine, with a cold wind blowing.

Greenwell's Glory Spider: This fly represents a variety of olives and is good throughout the season.

Yellow Partridge or Grey Gnat: A good killer during April.

Orange Partridge: An excellent killer from April to September on warm days.

Dotterel: A good standard fly all through the season from the end of April, more especially on rather cold days.

Poult Bloa: A fair killer on cold days throughout the season.

Knotted Midge: Does very well on hot, stuffy days when thunder is about.

Smoke Fly: This is a fancy fly and only kills in certain curious states of weather on sluggish water in dull, heavy, sultry conditions.

Black Gnat: Most difficult fly to imitate. Useful when fish are smutting.

Hackle Blue Upright: Probably one of the finest flies to use during the season and will often help to bring success on what would otherwise be a blank day. If there was only one fly allowed, that would be the choice, followed by Partridge and Orange and Greenwell.

Quite obviously, a good knowledge of entomology is essential for the serious fly fishing enthusiast. An instinctive take on the time of year and weather conditions that occur in certain fly hatches, with which to match an artificial, is not just the difference between success and failure, it can also allow the experienced angler to evaluate a whole season and find reasons why it may have been a good year or bad.

The dry summer of 1995 was awful for trout fishing and while 'low water' was an easy excuse for many of us who failed miserably that season, a more experienced analysis, such as that of Ken Smith of Durham City Angling Club, painted a picture with a far deeper perspective. In the autumn edition of the club newsletter, Ken wrote:

The feeding activity of trout is normally influenced by air pressure, rainfall and water temperature. Our weather for much of the season was dominated by high pressure, rainfall was low and normally the water temperature would have been quite high in the summer months. However when cold compensation water (from Kielder Reservoir) was introduced, from time to time, the river water temperature was varying very much – this factor has had a great influence on the natural food of the trout. Sedge hatches have been very poor this season and many believe it is as a direct result of the changing water conditions. Apart from sedges, there seems to have been a slight decrease in the number of up-winged flies, but there has certainly been an increase in the population of midges. This latter fly life requires silt for part of its life cycle and silt is now very evident on stretches of the river where once it was never seen.

Yet the experienced angler is able to turn such situations to his advantage and, in 1995, Ken Smith was no exception.

New conditions need new approaches if the angler is to continue to be successful. The main change that I have made, this season, has been to fish much finer than in the past. This was forced on me by the very clear water conditions and, therefore, the extra wariness of the trout. Fine fishing starts with the use of very fine cast nylon – 3 lb as a maximum. Large flies do not fish properly on fine nylon, so hook sizes were likewise reduced. Size 14 is now the largest used in clear water and 18's and 20's are very often needed for success. The use of tiny flies has certainly proved effective as the season progressed. Who would have thought of using midge pupae against river trout? Several anglers have this summer, with good results!

Of course, to justify making the game as hard as this, the clubs and associations (and even river boards, water authorities and the Environment Agency) had also to play a part in providing sufficient stocks of trout for anglers to fish for. By the start of the twentieth century, the stocking of trout streams subjected to regular angling pressure was standard practice, maintaining fish stocks and even increasing stock density in places where the burden was unduly heavy. The process was normally carried out annually by the burgeoning river fisheries boards, using juvenile trout reared in their own fish farms but, increasingly, angling clubs began to take matters into their own hands, using revenue from membership subscriptions to buy their own brood stock and, in some cases, even opening their own independent hatcheries.

There are very few records of these sorts of introductions readily available to public scrutiny, but thanks to the foresight of one North East angling club, there does exist a detailed record of their trout stocking and, as such, a blueprint for similar activities on spate rivers elsewhere in the region.

The aforementioned Derwent Angling Association, founded in 1865 as the Derwent Valley Angling Association, had as early as 1883 embarked upon a programme of restocking its river with juvenile trout and trout ova. Initially, the Association stocked both *Salmo levensis*, a strain of brown trout originally thought of as a separate species, and the truly unrelated *S. Fontinalis*, a strain of North American brook trout.

Early results were promising – at least in so far as catch returns for the 2,000 stocked 'loch leven trout' were concerned. Vanishingly few sightings were reported of their 4,000 North American cousins, however, and the DAA resolved from that point onward to stock only indigenous trout and grayling into the river – a policy that remained in force for almost the next one hundred years.

With the angling population still comparatively sparse, restocking was not repeated on the Derwent for another two decades, until, in 1901, the DAA invested £13, 10s in 1,000 *levensis* yearlings from Howieston Fisheries of Stirling. This number was repeated in 1907 and then, in 1908, the same amount was divided evenly into 500 *levensis* juveniles and 500 *fario* (brown trout); the two were still thought of as separate species.

Records also show that thought was being given as to where exactly to stock trout into the river, with stocks of the earliest fish the Association procured (supplied as eggs and fry) being introduced into tributaries close to the source of the river, where they might remain out of reach of adult fish and other predators. By 1908, however, the club archives record that the larger trout they were by then introducing were being placed into the main river throughout its fishable course, although *fario* tended to be stocked further upstream and *levensis* in the middle reaches.

By the end of the First World War, the DAA's policy of restocking the Derwent was well established, having continued on right throughout the four years of the conflict to satisfy the Association's commitment to provide free fishing for servicemen on leave. The two 'species' of brown trout were by now recognised as one and the same, and the practice of spreading the introduction of 1,000 yearling fish out evenly between the Association's upper boundary, near Blanchland, and its lower limit at Lintzford was repeated in each of the years 1920 to 1927.

By the early 1930s, the DAA had been joined in the act of restocking the river by the Tyne Fisheries Board, the forerunner of the later River Boards and Water Authorities, which gifted around five hundred trout to the Association – the latter placing almost half of the young fish in stew-ponds to 'grow them on'.

If this particular experiment had been of great delight to some members in subsequent seasons – given the sudden upturn in the incidence of 'specimen trout' catch reports – the concept of restocking, in general, was already

The River Derwent.

extremely popular in an era of virtually unlimited bag limits (the number of trout each member was allowed to kill in any one day). Nonetheless, the cost of the restocking was slowly creeping up – in 1930, the cost of a thousand yearlings from the Tyne & Eden Fish Hatcheries had been £22, while by 1949 the same number of fish cost almost double that amount.

To begin with, in order to alleviate angling pressure on the river, a limit was set on the number of members the Association allowed to join. Then, from 1960 onwards, bag limits started to be reduced – from twelve fish a day in the 1960s to six in the mid-1970s, with further reductions being made progressively, until just two trout were allowed to be killed per angler, per day by the 1990s – combined with a steady increase in the minimum size limit that was set for obtainable trout.

The Association even reprised the idea of stocking non-indigenous trout again in the 1970s and 1980s – rainbows on this occasion – and although this more hardy 'alien' featured more readily in contemporary members' catch returns, the experiment was discontinued. However, with a reservoir now situated in the Derwent's upper reaches, escapees from its commercial trout fishery continue to be taken in DAA waters to this day.

Nowadays, the cost of restocking trout streams is probably the greatest single limiting-factor to the finances of most North East angling clubs. In the

spring of 1996, the year after the drought so sorely tested Ken Smith's fly
fishing skills, Durham City Angling Club stocked 300 fish into the Shincliffe
stretch of the Wear, a figure that represented their annual introduction
of brown trout into club waters, thirty of which had weighed over 1½ lb,
according to club records. By late 1997, another quiet season was reported,
but there was an unexpected bonus in the shape of some fantastic spring sea
trout fishing above Shincliffe Hall. Like most North Eastern rivers, the Wear
was not then noted for a particularly good run of spring fish, but the news in
Autumn 1997 certainly showed signs of it bucking the trend.

'The start of the 1997 season fished quite well but, by the end of April,
things went quiet as far as brown trout went,' reported the club newsletter.
'However, those anglers who persevered soon discovered that sea trout were
pausing on the stretch on their way upstream. This was a welcome feature as
these fish have, in the past, rarely stopped in this water.'

The report continued, 'A swift change of tactics began to result in sea trout
being caught. They responded well to a biggish fly, fished deep, and just to
show their perverse nature, they also responded to surface flies, in particular a
muddler minnow. The best fish taken was said to be around 10 lb.'

The unpredictable nature of the fishing was even reversed when brown
trout began responding to sea trout tactics: 'Those anglers who had purposely
gone after sea trout also discovered that the brown trout were there for the
taking, as the deep lure intended for the sea trout was proving effective for the
brownies. By the time the sea trout had departed, this tactic was taking some
nice brown trout, but all good things come to an end and by late May the
sport tailed off.'

Nonetheless, the nature of that season's trout fishing was seen as indicative
of the fast improving situation with regard to sea trout on the Wear. The
Durham newsletter continued, 'One very interesting feature of this season has
been the sea trout. These fish have been running the river throughout the year,
even in the lowest of water. The old idea that sea trout only entered the river at
a certain time has not been the case this year.'

The runs of sea trout on both the Wear and the larger River Tyne were
interrupted during their respective 'pollution years' of the nineteenth and
twentieth century's, mining (for both coal and lead) and sewerage discharge
being the principal culprits in each case. The Wear averaged around 300
declared rod catches per year from the mid-1950s to early 1990s, a figure
that has more recently increased to about 1,500, with peaks of over 2,000 in
2002 and 2004. As was the case with its salmon, catches of sea trout from the
Tyne almost bottomed out in the three decades from 1952 to 1982, but the
combined total on the three main rivers of the system has now recovered to
a value similar to that of the Wear. Both major Northumbrian rivers are now
among the very finest for migratory trout angling in the UK.

Catch returns for the smaller River Coquet are more modest, but have remained fairly constant since the 1950s at around 300 to 400 sea trout per year. Even less famous rivers in Northumberland are gaining a share of the Northumbrian coastline's thriving migratory trout population, with the Wansbeck, in particular, seeing the benefits of fish pass installations on several of its many weirs.

Improvements in water quality as well as habitat management have paid dividends for all trout fisherman in recent years – those that go after brown trout and sea trout alike. It would be nice to think that in another ten to fifteen years, sea trout would once again ascend every Northumbrian river from the Tweed to the Esk in good number, just as the brown trout already proliferates throughout the region. This beautiful and interesting fish, to use Robert Blakey's words, marks the first chapter in many a North East angler's career. You might move on but you can never leave the trout behind; it is the beginning and the end of fishing in the region.

CHAPTER TWO

Four Rivers and a Fly Rod:
A Brief Guide to the Fishing on Four Northumbrian Trout Streams

Beginning or End, the question you are most likely to be asking by now, given that this is supposedly a fishing guide, is 'where do I start?' Of course we already know that the list of potential answers to this is lengthy, with every river in the area, from the Tweed down to the Ure, containing trout throughout most of its course. Therefore, to provide much more than a brief mention of each river, the locations and the contacts for getting permits and such, would be far too exhaustive. Nevertheless, no guide worth its salt could ever suggest it was complete without a detailed description of at least some of what is on offer.

This chapter is therefore a series of river walks describing the trout fishing on four North Eastern rivers, two of which are fairly small yet nonetheless highly prolific brown trout streams, and two are a little larger and feted for their brown trout and sea trout fishing alike. The descriptions take the form of a linear walk (some of which have interruptions in places) beginning at the upstream end of the waters of a particular club or association on each river, the Coquet, the Wear, the Derwent and the Wansbeck. The explanations will include, where appropriate, some idea of the techniques most appropriate to each beat, but as I can only claim to be an 'enthusiast' at any form of fishing, I'll assume that in most cases you'll be able to draw your own conclusions! The guide is intended to be exactly that – advice on how to get to and how to get through each fishery on foot, with places to park your car or, if you prefer, an explanation as to how to get there on the bus. By the end I hope to have answered the question 'where do I start?' not once but four times, and to have at least provided an introduction to the glories of some of the river trout fishing available in the region.

And there could be no better trout stream to begin with than the magnificent River Coquet. Rising high in the Cheviot Hills close to the Scottish border, this is the only proper spate river flowing directly to the sea that lies fully within the county of Northumberland – the rivers North Tyne, Rede, Allen, Breamish and Till are all tributaries, and the Blyth, Aln and Wansbeck are comparatively small and truncated.

Effectively cutting Northumbria's largest county in two, the Coquet has arguably carved out the most spectacular valley in the whole North East; and it ticks all the boxes for the dedicated trout angler. Its large mountain-fringed catchment provides a continuous model habitat for the indigenous brown trout population – once a totally unique strain – and ensures regular spate conditions to draw a substantial annual run of large sea trout into the river and to help keep them on the move. The distinctive strain of brown trout has now sadly all but disappeared, its bloodstock diluted to the point of no return by generations of farm-fish, but at least, for many decades, the stock trout were reared on their parent river at the hatchery of the main angling club near Rothbury.

The Northumbrian Anglers' Federation took the lease for much of the fishing on the Coquet as long ago as 1897, and still to this day controls over ten miles of river between Thropton, at the bottom of Coquetdale, and the estuary between Warkworth and Amble. The first significant stretch of the Federation's water occurs around Rothbury – essentially the middle reaches – beginning just upstream of the main village in the Coquet valley and continuing down to the hamlet of Pauperhaugh, approximately four miles downstream.

There are many fine pools on the Coquet that will appeal to both salmon and sea trout anglers alike – the sea trout run being predominantly in the autumn – but for the brown trout angler there is greater freedom to roam, with most of the river amenable to this branch of the sport. The NAF waters at Rothbury begin at Lady's Bridge, a footbridge easily accessed to the west of the village (park in the village, head out along the road towards Thropton and after about half a mile, take the first left turn down a track towards the river), with the imposing shadow of the sandstone Simonside ridge looming up immediately to the south. The river initially runs across the local golf course and is relatively lacking in bankside features, although the pools themselves provide ample habitat. A footpath on the left bank (as you look downstream) follows the river all the way back into Rothbury, although fishing is permitted on both sides.

Still a good fifteen miles from the sea as the crow flies, by this point the Coquet has already acquired the status of a medium-sized spate river, with a riverbed that frequently includes a much wider expanse of parched stones and gravel in drier weather. On arrival in Rothbury, the river is channelled through a much narrower gap than the wide flood meadow it passes over just upstream and it gathers pace as it rushes beneath the main road bridge carrying the B6342 from Scots Gap. This road comprises the best route by car to Rothbury from the Hexham/Tyne Valley area, although the best way to come from any other part of the North East would be via the well-signed A697 and B6344.

The footpath, still following the left bank, is likely to be much busier now, with tourists of the human variety as well as the silver ones in the water, as the river at last gains some bankside feature in the form of trees, with a caravan park on the opposite side. The river meanders gently to the right and left and then, more sharply, to the right again with one or two pools worth a cast – although watch your back cast here if you're fly fishing! The Coquet's brown trout respond well to many fly patterns – early season, weighted nymphs such as the Pheasant Tail and Gold Bead Hare's Ear, as well as classic wet fly patterns like the Greenwell's and March Brown will score, while later on dry olive and mayfly patterns will come into their own. The Partridge and Orange and the Snipe and Purple will work in either wet or dry fly forms for much of the season.

As the river leaves Rothbury, it approaches a sharp right-hand bend at a limestone rock outcrop beside local landmark Thrum Mill. The river narrows and gains pace markedly, as it cuts through the rock to form the impressive Thrum Mill Falls, yet it was once even narrower here and leaping across the treacherous current used to be a common pastime until, in the eighteenth century, a young boy fell in and was drowned. The falls were subsequently blasted with dynamite to both widen the gap and suppress the current, and while the pool below the falls is still, needless to say, a very useful place for fishing (check the rules for restrictions at certain times of the year), the swift waters immediately above can also be worth a go.

From Thrum Mill downstream to Pauperhaugh, the Coquet is a constant series of glides and deep pools running through open countryside next to the B6344, whose numerous verges, which are suitable for parking, make this an ideal beat for the night-time sea trout angler. For the daytime fisherman, parking in Pauperhaugh itself is best, by turning into the minor road that's on your right (if you're heading away from Rothbury) and parking by the roadside on the far side of the bridge. The river twists and turns throughout this section without ever leaving its predominant easterly direction and this continuum of Federation water finally ends about half a mile below the bridge at Pauperhaugh.

There now follows a 9-mile interruption to angling available to the general public, although there is salmon and trout fishing offered by day ticket and to guests of the Anglers Arms at Weldon Bridge, where the B6344 meets the A697. By the time NAF fishing resumes on the right bank, just upstream of the two bridges in the village of Felton, the river has matured and the moorland that still fringed the valley top above Pauperhaugh has given way to pastoral farmland, with the occasional open-cast mine.

Felton is easy to get to – situated on an old section of the A1, 10 miles north of Morpeth, which was bypassed back in the early 1980s. Parking is possible beside the river where this lower Federation beat commences and access downstream is by means of a well-defined public footpath that follows the right bank all the way. In the village itself, fishing is good both upstream of and between the two bridges, one of which is described by Robert Blakey in his story of the country lad and his big trout in 1826. Immediately downstream, however, the river becomes a very long, shallow glide, absolutely no use for fishing whatsoever, which slips between densely wooded banks for several hundred yards.

By the time the downstream path has come back alongside the river, the rapids have broken, the woodland disappears and there is a succession of pools well suited to fly fishing as the river runs through open countryside with sparse bankside cover for a mile and a half down to the mouth of a small tributary, the Thirston Burn. The burn is easily crossed to continue following the main river, which now swings sharply to the left to continue in a northerly direction for just over a mile. The Coquet here is becoming similar in nature to the larger rivers Wear and North Tyne; its frequent winter spates have carved high banks into the relatively flat valley floor at every turn, and the riverbed is wide with dry gravel margins and islands appear in places. There are several excellent holding pools for sea trout on this stretch, with access to the water from the dried-up sections of riverbed that are relatively easy at night.

As the river begins to turn back to the right, the valley quickly narrows with the river gradually deepening as it heads downstream. Access continues along the right-hand bank, but the fishing rights, which have hitherto been from either side, are now restricted to the bank with the footpath. The river again swings to the left soon after and the open and relatively flat surroundings that have accompanied us all the way from Felton give way to a steep, densely wooded bank side. This slow deep section, heavily featured and known locally as Sandy Hole, is all but un-fishable by fly, but it does have a fine reputation for producing some very big brown trout to both worm and spinner. This marks a fitting end to trout fishing on the Northumbrian Anglers' Federation waters on the Coquet. The boundary sign is soon upon us and although there is another section of mainly two-bank fishing beginning approximately five miles downstream at Morwick Mill, this is subject to special rules and essentially preserved for salmon fishing only.

As described, getting to Rothbury and Felton by car is relatively easy. Public transport to these fairly remote locations is less straightforward, however, although Felton is served by hourly bus services (Arriva, 505 and 501) to and from Newcastle Haymarket Bus Station, Morpeth and Alnwick. To get to Rothbury by bus it is necessary to travel to Morpeth Bus Station first and then connect with the local 144 service, which continues to Thropton.

The Northumbrian Anglers' Federation offer season permits for trout (only) at a very reasonable price or, alternatively, trout and salmon fishing can be paid for at greater cost. Permits are available by writing to the Head Bailiff, 15 Woodlands, Rothbury, Morpeth, Northumberland, NE65 7XZ, telephone 01669 620984. The NAF's website is: www.northumbriananglersfed.co.uk, and trout permits can also be purchased from McDermott's fishing tackle shop, 112 Station Road, Ashington, Northumberland, NE63 8HE, telephone 01670 812214.

* * *

While the Northumbrian Anglers' Federation holds almost all of the publicly available fishing on the Coquet, down on the River Wear there are any number of clubs and associations you can join to gain access, with fishing available from high up in Weardale to the tidal reaches downstream of Chester-le-Street. The Wear is the third largest spate river in the Northumbria region and as a fishery, in its lower reaches at least, it is best described as a mixed fishery. Suffice to say, brown trout are prevalent throughout the river and its two small tributaries, the Browney and Gaunless. So too are sea trout, and as we already know, the Wear is regarded by many as the finest sea trout river in England.

For our fairly short walk along the banks of this excellent river we have come to its middle reaches and the waters of Durham City Angling Club. Established in 1962, DCAC is one of the few clubs on the river that caters for coarse and game anglers alike. As such, a considerable portion of the club's 6 miles of bank between Croxdale and Chester Moor is best suited to fishing for the river's many coarse species. Nonetheless, DCAC's game anglers are well looked after and their favoured locations are found on those club waters upstream of Durham City.

To start with, we are visiting the mile-long beat at Croxdale, a fishery that runs down the left bank upstream and down of the main A167 Durham to Darlington Road, and is shared with Ferryhill Angling Club. Access by car is obviously easy and parking is allowed in the car park of the Honest Lawyer pub, about two hundred yards north of the main bridge over the river. Alternatively, you can leave your car quite securely by the roadside at the southern end of the old stone road bridge that once carried the A1. Access is by turning into the B6300 opposite the Honest Lawyer and taking the first left turn after about a hundred yards.

The beat is divided into two parts, the upstream section being only about four hundred yards long and bounded at each end by the old road bridge and an impressive viaduct, which still carries the main east coast rail line. The upper section, like so many on the middle Wear, is actually better suited to coarse fishing tactics than to fly fishing, but a worm or (after 16 June) a maggot is still more likely to catch a trout on this section.

The beat commences where a glide breaks beneath the tall arches of the viaduct, the water quickly becoming deeper and the current more sedate. The fishing is mainly from 'pegs' cut into the bank side – again in the coarse fishing

A River Wear sea trout.

style – with the bank fringed by tall trees, some of which crane out over the river to provide cover for the fish. The best spot to fish – if you can get there first – is a double peg right in the middle of the upper section, bang in front of which is a depression or 'hole' in the riverbed that is around 12 feet deep. A worm fished from here hard on the bottom just downstream of the peg will often score and I once landed a very early season sea trout of about 4 lb in this way.

The nature of this swim makes it of interest to non-human anglers as well, which may just make an appearance if you keep very still. Another time, I was sitting waiting for a bite when out of the corner of my eye I saw something black and about the size of a large cat swimming down the middle of the river, only yards from where I was sitting. I knew it was too large to be a mink, a species which had wreaked havoc on the local wildlife in this area for many years, and as it passed me without so much as a glance, I could see it was an otter. It continued another ten yards on from where I had cast, to the spot where it thought was best for fishing and executed a textbook dive; the last thing I saw was the curled tail and back legs of the animal as it made for the bottom of the pool. I didn't catch anything that evening, but I didn't really care!

The downstream section of the Croxdale beat begins on the same bank immediately below the new road bridge after an interruption of about a hundred yards where the Wear cascades over boulders between the two bridges. The river continues to descend for a fairly long glide following the falls, passing beneath what looks like a disused army rope-slide connecting the two banks, before breaking into the first pool about two hundred yards below the bridge. With bankside cover becoming increasingly sparse the further downstream you go, this is now fly fishing territory with both brown and sea trout tactics bearing fruit in this first fairly shallow depression in the riverbed.

Another short set of rapids results in another slightly deeper pool, which is again productive for trout. On the Wear, both brown and sea trout will respond well to the patterns described by J. L. Hardy in the previous chapter, as well as to the Gold Bead Hare's Ear, Pheasant Tail Nymph, Watson's Fancy, Dunkeld, Grey Duster, Black Pennel and Bibio. It's also worth noting that this stretch is one of the more productive areas on the river for grayling, which will take many patterns intended for trout but also Czech nymphs, grayling bugs and klinkhammers fished upstream.

The second pool breaks into 200-yard-long stretch of swift running, fairly shallow water that never quite becomes rapids and retains sufficient depth throughout most of its course to be fished for brown trout with a wet fly. This then breaks into the last pool on the beat, a very long (about 500 yards) body of water that becomes, to all intents and purposes, a coarse fishing stretch at its much deeper and slower bottom end.

At the top of the pool, however, is quite a useful run where the fast water enters, producing an almost diagonal flow from right to left that is ideal for conventional wet fly fishing across-and-down. In conditions of normal flow, the river is shallow enough to be crossed at the head of this pool and, although fishing from the far bank is not allowed, casting from a position fairly close to that side is the best approach. The riverbed is gravelly here, and gently slopes diagonally downstream from right bank to left, meaning that you can get into a very favourable position for casting by wading carefully forwards. There is also a deep run on the left bank here where fast water gurgles beneath an old tree that fell into the river after floods in the late 1990s. This makes for an interesting place to try a cast upstream and is often a lie favoured by grayling.

The Croxdale beat ends where the Wear meets the smaller River Browney at the bottom of the large triangular field that has bordered the river all the way down from the road bridge. However, those fishing on a Ferryhill ticket can continue on downstream of the confluence as their club controls the left bank (Brock Bank Farm) for a further mile downstream. For Durham members, there is now a 2-mile interruption as the river meanders over the wide flood plain on its course towards the ancient cathedral city.

Before reaching the outskirts of Durham City, the Wear passes the village of Shincliffe which stands on the right bank less than a mile from the city centre. Not far from here is Shincliffe Hall, which is reached by turning into the village off the A177 beside the Rose Tree pub and then turning right again (essentially turning 'straight ahead' as the main road swings to the left) and following the narrow lane past a garden centre and on between two fields. After a quarter of a mile the lane goes uphill slightly as it reaches a stand of trees, then drops back down into woodland just before you reach the Hall buildings. On the left are the designated parking spaces for anglers fishing DCAC's Shincliffe Hall beat, another excellent stretch for trout fishing.

Shincliffe Hall stands at the bottom end of the beat, so walking to the top end about half a mile away is by either following the river (take the path to the right after passing the hall) or by following the main footpath, which cuts through the edge of woods to the left of the riverside meadow before coming back alongside the river a little way upstream. Where the path, river and woodland all re-converge, DCAC's Shincliffe Hall stretch begins on the right bank, a constant series of pools fed by fast water that are only lightly featured on both sides.

This is the only stretch of river under DCAC's control where bait restrictions apply over and above the normal Environment Agency seasonal byelaws. Elsewhere, the restriction on using baits other than worm would be lifted from 16 June, but at Shincliffe Hall only bread can be added to the list before the trout season ends on 30 September. This effectively preserves the beat as a trout fishing venue, encouraging summer coarse anglers to make use of those sections of the river more amenable to that type of fishing further downstream.

At first, the banks are very high and steep but once you get away from the woods, while never completely straightforward, the water does become more easily accessible. The path running along the top of the bank makes the journey from pool to pool quite easy, however, and being on the inside of a long, right-hand bend, most consist of swift runs of water down the far bank with the margins mainly shallow enough to wade; the flow only belatedly strays over to the near bank towards the bottom end.

Although a good area for brown trout, this is an extremely popular place for sea trout fishing and anglers will usually be seen either parking or walking the bank in the hours approaching dusk from April through to September. And although bait restrictions lift and coarse anglers become more frequent on the half-mile stretch below the Hall down to Shincliffe Bridge, if anything, the holding pools become better and sea trout are often seen jumping as light fades in the deep hole immediately upstream of the bridge – always worth a look!

Durham's water continues for another mile adjacent to, and opposite, Maiden Castle, which is situated on the left bank below the bridge, but we are now entering those areas better suited to coarse angling and the best of the trout fishing is behind us.

Needless to say, Durham is well served by public transport and Shincliffe village is just over a mile's walk from the city's bus and rail stations. However, the 56, X1, X2 and X41 Arriva services all go through the village from the city centre. For Croxdale, which is about three miles outside Durham City, several buses stop right outside the Honest Lawyer, including Arriva's X24 service from the Metro Centre to Bishop Auckland, which calls at Durham, and Go North East's 21 service, which goes from Newcastle Eldon Square Bus Station to Bishop Auckland, again via Durham.

Details on how to join Durham City Angling Club are available from their website, www.durhamanglers.co.uk, or by writing to the Membership Secretary, Durham City Angling Club, PO Box 508, Durham City, DH1 9BP.

***

Mention the word Derwent and most anglers' minds will turn to that iconic Derbyshire trout stream, or to Yorkshire's excellent coarse river of the same name. Non-anglers will most likely think of the Lake District and Derwent Water, some realising that this famous lake is both fed and drained by another River Derwent – itself an excellent trout stream.

Even in the North East, most anglers' first thoughts will be of Derwent Reservoir and its hallowed reputation as a commercial trout fishery. Yet it seems few even realise that there is a fourth, far less famous River Derwent, the one from which the reservoir takes its name, a river over 30 miles in length that also forms a considerable part of the border between Northumberland and County Durham.

This Derwent rises above the Northumberland village of Blanchland, flowing east before discharging into the reservoir at Ruffside. On exiting below the dam, the river runs south-east to Eddis Bridge where it accepts releases from two other much smaller reservoirs, Hisehope and Smiddy Shaw, via the Hisehope Burn. From here, the Derwent settles into a more or less continuous north-easterly direction, passing the villages of Allensford, Shotley Bridge, Ebchester, Blackhall Mill and Lintzford. By this point, it has acquired the runoff from four significant tributary streams and is of a similar width to the Coquet at Rothbury, a size it retains for the next ten or so miles to its confluence with the tidal River Tyne, near the Metro Centre.

For angling purposes, the lower river, from Lintzford down to its tidal reaches, is preserved more or less continuously by the Axwell Park & Derwent Valley Angling Association, an angling club that takes its name from an old colliery of the same name that used to work the bottom end of the valley. APDVAA was founded in 1928, presumably by miners that worked at the Axwell Park Pit, renting banks on the river around the nearby villages of Winlaton Mill, Lockhaugh and Rowlands Gill.

The Derwent is, by nature, a trout stream. Except for its short tidal stretch, it lacks any of the coarse species found in the River Tyne and its grayling stocks are merely the result of introductions at around the turn of the twentieth century. Yet unlike angling clubs on certain other North East rivers, the APDVAA (and its upstream neighbour, the Derwent Angling Association) never regarded its grayling as anything other than a blessing; there were no heavy-handed attempts to rid the river of an unwanted 'alien coarse species' here! With the more delicate thyllamids protected by both associations through bait restrictions and catch and release rules, rather than a trout stream, the Derwent is controlled in the manner of a game river with two species – the indigenous trout as well as its graceful salmonid cousin.

Indeed, there isn't much on the Derwent that might appeal to any angler without a fairly traditional approach to the fundamental principles of game fishing. The DAA totally prohibits the use of any kind of fixed spool reel, and while Axwell Park does permit worm fishing after 1 June, this is only on those sections of the river adjoining the parks at Rowlands Gill and Derwenthaugh. The rest of the river remains fly-only all season long and few members are ever seen carrying bait rods.

The APDVAA's water commences not far below the weir at Lintzford, about half a mile upstream of Rowlands Gill. This stretch runs on the left bank though a deep wooded gorge alongside the A694 Shotley Bridge road and is the haunt of big trout and specimen grayling in its darkened waters. Parking is available on a redundant section of old road alongside the A694 about a hundred yards the other side of Lintzford, and the inaccessible nature of this half-mile beat's numerous deep pools certainly lends itself to the better quality of its inhabitants. Be warned, however, the rocks can be treacherous and the pools deepen very sharply in places, so take extra care and maybe even a wading stick on this section.

The Lintzford beat lasts about half a mile to the edge of Rowlands Gill, and after a half-mile break, because of bankside housing, the fishing resumes on that side beneath an old railway viaduct that now carries the Derwent Walk country path from Gateshead to Consett, in addition to the Coast to Coat Cycleway from the Tyne Piers to St Bee's Head in Cumbria. The river makes a wide left-hand turn here, creating an extensive far bank deep that can be productive to a well-presented fly, despite the sedate nature of the flow.

The slow current is caused by a small weir situated on the downstream edge of a road arch about a hundred yards below the viaduct, the bridge carrying the B6314 to Burnopfield. It is at this point that Axwell Park's water commences on the right-hand bank, at the beginning of land belonging to the National Trust's Gibside Estate. Parking here is free, not in Gibside, but in a rubble car park just off the B-road next to the start of the Derwent Walk on the Rowlands Gill side.

From the bridge, the water on the Rowlands Gill bank is available to fairly inexpensive day tickets that can be bought at the caravan park shop about

The River Derwent at Rowlands Gill.

quarter of a mile downstream. The far bank (Gibside) is members-only, although day tickets that allow access to all the association's waters can be obtained at a greater cost.

The first feature is the weir pool, a swim best fished from the right bank, not least because of the fact that the park side is often overrun with tourists in the warmers months – including those with a tendency to paddle and throw sticks for dogs to chase into the water. Consequently, the best time to fish here is either early or late but, with the bank on the estate side extremely steep, if you do decide to fish during the day, you should at least have that bank all to yourself.

Which is just as well, as the other consideration is that the only way to fly fish effectively is off that side. In conditions of normal flow, the current running over the obstruction passes through the arch nearest to the right bank, flowing diagonally beyond the crest of the weir before forming a back-eddy that slides serenely past rocks on the park side that belie the sudden depths the pool plumbs.

The depth on the fishable side is quite obvious – and sudden – with a brisk current lasting twenty to thirty yards even in low-water conditions. Close to the weir, the pool fishes well to a weighted fly presented upstream in the Czech Nymph style, with grayling being the species most often caught in this way.

Try the GBHE or Pheasant Tail in tandem with a March Brown early season, with the brown replaced by a black pattern later on. Towards the tail, with the depth decreasing slightly and current subsiding, 'across-and-down' in the traditional wet fly mode is best and, of course, the whole feature can be fished – at the right time of year – with a dry. Various olives or a grey duster might work best in this instance.

Downstream of the weir pool, the river becomes a quick succession of glides and shallow pools until it starts to deepen ahead of a sharpening right-hand bend downstream of the caravan park. The river begins to meander quite markedly here, retaining a reasonable depth while sharp bends produce several tantalising runs. Be careful here, though, as electric wires cross the river at several points, including one of the better looking spots!

Access to this part of the river can be from either bank, although wading across is impossible for some distance, owing to the depth. The right bank can be followed by walking up the access road for Gibside before climbing over a gate into an open field. You can then follow the fence on your left that marks the top of the bank, climbing through at several points as the river swings round to the right.

The left bank is less straightforward here. Access from Rowlands Gill park is easy but after the caravan site, right of entry is blocked due to a conservation area and it's a fair walk from here to get to the next section of bank (an annexe) on that side. Firstly, follow the main A694 towards Newcastle for about quarter of a mile, then bear right onto the Derwent Walk. Follow this path for another half a mile, then bear right again through a gate, at an old railway bridge, before taking the metalled lane to the right down a long hill. This access road leads to a car park for bird watchers and access to the river is possible by following a path past a small pond with a bird hide.

While the left bank is interrupted again by another conservation area, the opposite bank is continuous right through Gibside, although deviations away from the river may be necessary in places due to the rough nature of the terrain. There are paths, however, to achieve a complete passage through to Lockhaugh Meadows, with several highly productive runs en route, including 'the wall' where the main footpath comes right alongside the river. Soon the Derwent, which has been almost straight for about a mile, begins to meander again as it hits a succession of rocky outcrops, which in turn produce several highly productive pools.

By this point we have reached Lockhaugh Meadows, the fishing on the left bank has resumed, and the river must be forded if following it downstream from the Gibside estate, this being possible at a set of rapids immediately down from where the river begins to swing round to the right. The Gibside beat quickly becomes inaccessible due to high banks, as the Derwent snakes first right and then left in a sweeping 180-degree turn that lasts little over a quarter of a mile, before passing back under the former railway at the majestic Nine Arches Viaduct. The bankside foliage, which has been intermittent on the straight section, becomes far denser here but a fly can still be placed over most lies with careful casting.

A gently undulating open field on the inside of the bend affords easy access from the left bank, the first big pool commencing where the river rushes over a limestone sill at the bottom of the rapids where you cross, forming what is almost a natural weir pool. This feature is around 40 yards long and 8–10 feet deep in places, courtesy of the relentless rush of water. By the bottom end of the pool, the river is already being pushed back to the left by the towering right-hand bank, making the whole run an ideal place for casting wet flies and nymphs across-and-down. At the tail, the current is forced over another set of rapids and arrives at the next much shallower pool after rushing through another limestone outcrop. This pool is almost the same length as its upstream neighbour but is almost straight and, because of this, its depth and stronger current is less prolific, although it is always the home of a good number of trout.

Once again the river rushes through rocks, with deep pools containing fast water that will fish to a heavy fly as it swings left in a relentless arc beneath the high banks on the far side. At last, about two hundred yards upstream of the viaduct, the river enters a more sedate phase, lunging under one last far-bank limestone shelf before slowing into a couple of slower shallower pools in the shadow of the nine arches.

The fishing ceases here on both banks for about quarter of a mile and access into what is the park (formed from land on which the former Derwenthaugh Coke Works stood) is by ascending the steps to the left of the river and crossing over the viaduct. Immediately on the other side, a metalled path on the left (signed as the C2C cycleway) drops back down to the river, the point at which it comes alongside the right-hand bank being the place where fishing recommences. The river here is swinging back round in the opposite direction to its course on the upstream side of the viaduct, the tall hill on the inside of the bend being what remains of the coke works' former slag heap. There are several good pools here before the river slows on approach to a small weir called High Dam, an obstruction which used to divert service water to both the coke ovens and their predecessor, Crowley's Iron Works.

This weir pool is much shallower than both the one at Rowlands Gill and the top pool in the meander at Lockhaugh Meadows, but nevertheless it will fish. Just get there early! However, there are more productive pools just downstream which are far less accessible to both kids and tourists.

The first is a pool just fifty yards below the weir, over which a new footbridge was built to connect both sides of the park just before it opened in 1999. This pool is right underneath the arch and can be approached from either side, but casting upstream means you will probably be both out of sight and out of mind of the local teenagers that like to gather beside the weir on summer evenings! The river then runs another hundred and fifty yards – through an area that can be quite productive for grayling – before a sharp left-hand bend marks the end of the sweeping double meander, taking it back into its predominant north-easterly trajectory.

The fishing from here on down is mostly from both banks, the left-hand side adjacent to the village of Winlaton Mill being Derwenthaugh Park, and the right bank consisting mainly of open fields with a couple of brief interruptions. Access here is relatively easy on both sides, the only hazard being a row of electricity pylons whose wires cross the river about three quarters of a mile below High Dam. However, there is less bankside foliage on those parts of the right bank where you can fish and you should, in theory, be able to progress from pool to pool unimpeded by park-goers. Unfortunately, crossing the river on this beat has been made extremely difficult of late owing to the collapse, during the September 2008 floods, of the Butterfly footbridge that used carry the Clockburn Lonnen path coming down from Whickham. It is therefore necessary to choose a bank and stick with it.

Shortly after the pylons, the river begins to deepen markedly on its approach to another weir – the largest on APDVAA's waters – a feature beside a tennis court that marks the tidal limit of the river on its downstream side. The weir pool here is surprisingly small for such a large obstruction and not particularly deep, but it is nonetheless popular with anglers as, at certain times of the year, this marks the furthest point that sea trout taking a detour from the Tyne can ascend the Derwent. The installation of a fish ladder in the not-too-distant future should hopefully alleviate this situation and the prospect of sea trout fishing on late summer evenings from the river at Lockhaugh, Gibside and Rowlands Gill is a tantalising thought.

We're near the downstream limit of Axwell Park waters now, and a bridge two hundred yards below the weir allows the river to be crossed again and both banks of this tidal section are available to members for another quarter of a mile down to the B6317 Swalwell road bridge. It's been an 8-mile walk if you've been following the river all the way from Lintzford, but hopefully those trout and grayling have made it well worthwhile.

Regarding bus services, the 45 and 46 Newcastle Eldon Square Bus Station to Consett services pass through Swalwell, Winlaton Mill, Rowlands Gill and Lintzford, while the 47 Eldon Square to Blackhall Mill calls at Swalwell, Winlaton Mill and Rowlands Gill. All three services are operated by Go North East and call at the Metro Centre Interchange. In addition to the free parking at Lintzford and Rowlands Gill, you can also leave your car at Thornley Woods Country Park (on the A694 between Rowlands Gill and Winlaton Mill), which is about a half-mile walk from Lockhaugh Meadows, and there are two car parks in Derwenthaugh Park (Winlaton Mill) which are right next to the river.

Details on how to join Axwell Park & Derwent Valley AA or the availability, whereabouts and cost of day tickets can be found on their website: www.apdvaa.co.uk.

* * *

All three of the rivers I've mentioned so far hold a place in my own personal affections, but it's the River Wansbeck, for reasons that will soon become apparent, that's my particular favourite. The Wansbeck isn't big and it certainly isn't famous, but it is actually quite a good trout river, which is just as well, as there isn't much else in there besides minnows and eels! Sea trout have been recorded in recent years, though not in sufficient numbers to be taken seriously. Escapees from commercial trout fisheries further up the valley also figure in annual catch returns, but there are no coarse fish or grayling and the last rod-caught salmon was taken way back in the early 1960s.

Rising on moorland scarcely 20 miles west of where it runs into the North Sea, within a few miles of its source, the infant stream runs through two artificial lakes called Sweethope Loughs. Created in the late eighteenth century by the famous landscape gardener Capability Brown, this became the first put-and-take commercial trout fishery to be established in the Wansbeck catchment. Flowing out of Sweethope in an almost continuous easterly direction, the Wansbeck then runs through the grounds of the National Trust's Belsay Hall before continuing on to meet its first major tributary, the Hart Burn, a few miles further downstream. In fact, much of the river runs through private land and thus fishing has never been available, except to the occasional syndicate, on any part of the Wansbeck upstream of Morpeth.

The same can be said of its second significant tributary, the River Font, which converges with the main river at Mitford. The Font rises on the southern edge of the Simonside Hills and includes another locally famous commercial fishery, Fontburn Reservoir, in its headwaters. The confluence of the Font with the main river makes the Wansbeck downstream from Mitford a more substantial flow, although on those parts round Morpeth where the river has been dammed by weirs, its width can be somewhat exaggerated.

Meandering spectacularly beneath the tall valley sides in the market town of Morpeth, the river descends four weirs in the 3 miles between the mouth of the Font and the downstream edge of town, then two more in the 5-mile chicane between Whorall Bank and the tidal limit at Shepwash. Indeed, it is the sheer scale of these obstacles that's the main reason why the Wansbeck has never had a serious reputation for any form of sea trout or salmon fishing.

But for brown trout, admittedly mostly average in size, it does have an excellent name. The fishing from the upstream edge of Morpeth down as far as Sheepwash is, except for a mile-long stretch of free water on the 'town section', more or less all controlled by the Wansbeck Angling Association, a club whose roots are ingrained in Northumberland's most vibrant market town.

At its upstream end, WAA's water commences on the left bank in an area of woodland open to the public called Scot's Gill. Access is straightforward, with a footpath running adjacent to the riverbank all the way, and the stretch is easy to get to both on foot from Morpeth town centre (about three quarters of a mile) or by car, with a free public car park situated at the entrance from the B6343 Cambo road, next to Morpeth Rugby Club.

Scot's Gill begins at a moderately deep pool where the river runs over a limestone outcrop, a location that nearly always contains one or two decent-sized trout. The river then cascades down a fairly long glide before accepting the outfall of a mill race and settling into a slightly longer section of gradually deepening water. The trees here are very close to the water's edge and hence casting on this fly-only section has to be conducted with extreme diligence. Nevertheless, the rewards to the careful angler can be exceptional given the extent of cover afforded to the trout. And while the job of actually getting a fly onto the water may be quite demanding all the way down this beat, the job of fly selection is at least less of a challenge – most of the patterns recommended by J. L. Hardy in the previous chapter will still work here, in addition to, at the right time of the season, the March Brown and the Iron Blue Dun.

After the deep section there is another short glide – where the river can be forded from the main road on the far bank – which then breaks into another long pool that never quite plumbs the depths of its upstream neighbour. A slight advantage is that it's less overgrown, making casting less tricky, although this combined with the shallower depth leaves you with the feeling that there aren't likely to be as many monsters lurking here.

Again, the long pool gives way to a shallow glide after a few hundred yards, with this one breaking into another shorter pool, with the current slipping more or less straight down the middle. Again, this pool is amenable to either a dry or a wet fly, although heavily weighted nymphs are only really necessary at the start of the season.

The river now becomes a more swiftly running stream broken by boulders, before it slows and deepens slightly on its approach to an old stone road bridge (the B6343), marking the first interruption of Association bank.

There follows a five hundred yard break on both banks – an area of strictly private fishing that was once known as 'Matty's', after the local landowner. Access to the next short stretch of Association bank is by crossing the bridge and heading downstream – the club water continuing here on the right bank. Access is far trickier, however, with towering banks making entry something of a scramble unless you take a sneaky short cut by plodding through the margins of the shallowing downstream section of Matty's.

Unless you're an advocate of fishing a very heavily weighted nymph in very slowly moving deep water, only the top end of this long pool will be of interest to the fly angler. The rapids descending from Matty's deepen markedly after about twenty yards and the lower section of the beat – up to 15 feet deep in places is far better suited to fishing a worm, a method that is permitted on this section after 1 June.

Downstream from the second section of Association bank, a stretch of free water (water fishable in season by any angler in possession of a valid rod licence) commences on the same side where a footpath called Lady's Walk runs along the top of the bank to form a sort of promenade with a metal rail. The water here is still fairly deep and slow moving, so bait-fishing is again the preferred technique, although, unlike WAA waters, a worm can be used throughout the season, with any bait permitted after 1 June – bread paste being a particular local favourite.

Lady's Walk soon comes to an end beside a footbridge called the Skinnery Bridge – the first such structure in the area to have had a gas light, back in the nineteenth century, which was lit by fumes from the sewage pipe it originally carried beneath the footway. The river continues to be slow and deep as it starts to swing to the right with the high left bank occupied by private gardens that tumble down to the water's edge.

Free fishing is still available on the right bank that now borders a widening flat grassy field called High Stanners with the first major housing estate of Morpeth a couple of hundred yards beyond. The bend consists of slow deep water and a float-fished bait cast over towards the far bank can often be effective. As the river achieves a 90-degree deviation from its predominant easterly course, it begins to shallow up again and a set of stepping stones cross from High Stanners connecting to the main street in Morpeth, which is now only a few hundred yards away.

The free fishing is now from both banks and, if you don't mind a scramble and the constant attentions of exploring kids during the summer months, the left bank immediately up and downstream of the steps can be worthy of investigation. Below the steps the Wansbeck becomes very shallow again and the next set of rapids feeds one of the few pools on the 'town section' that can be fished effectively with a fly. After this, although fishing is permitted from both banks, for the next half mile after a more modern sandstone road bridge only the most ardent enthusiast would consider a cast as this is Morpeth's main park and both tourists and rowing boats abound on this slow-running stretch throughout much of the season.

By the time the river runs over another weir at the downstream end of the park, it has turned 90 degrees to the left and is now once again heading east. Immediately downstream of the weir, the water is very shallow, but on approach to another footbridge it does deepen slightly and the easy access

The River Wansbeck near Morpeth.

from this river crossing to both banks affords the fly angler another chance to practise his art.

Immediately downstream there is another more impressive stone road bridge, marked on its northern side by the towering steeple of St George's United Reformed church. This is the famous Thomas Telford Bridge, which for many decades carried the A1 into Morpeth town centre, and downstream of here the river runs over another set of rapids before deepening slightly alongside a concrete flood wall. Once again the Wansbeck begins to deviate to its left, guided for most of the way by the flood wall before it fizzles out at the end of the housing estate it was built to protect. The completion of this third right-angle bend in under a mile marks the third and last place that can be fly fished on the free stretch – a shingle beach on the left bank which is just upstream of a metal footbridge.

This beat can be approached by either crossing the bridge from the 'flood wall' side or from the town centre by walking down Gas House Lane past Morpeth Library. There is also ample car parking space at both ends of Gas House Lane, although parking in these town-centre locations is not free!

Below the metal footbridge, Wansbeck Angling Association bank recommences on the right-hand side of the river, which deepens again and bends sharply back into an easterly direction, faced on the left by the last few hundred yards of 'free' bank. Fishing here is again more amenable to bait fishing tactics, with the overhanging branches on the 'free' side immediately below the sharp bend often a good place to trundle a float.

The river is deepening on approach to another weir about quarter of a mile below the footbridge, this one having been built to provide water to power the mill buildings which still stand on the left bank. Below the weir, the Association's waters continue on the right bank, with the river now becoming more similar in nature to Scot's Gill, with a constant series of glides and pools.

The weir pool is itself is worth a few casts, with either a wet fly or nymph from a safe standpoint on the weir sill, or with a dry from the tail of the pool. A worm will score here also, with the excitement of watching a float trundling towards the bottom end of the pool palpable every time its progress is interrupted!

For the next half mile, the Wansbeck runs through open countryside, twisting sharply first left and then right before disappearing into the dense undergrowth of Quarry Woods and settling into a south-easterly direction for another half mile. This woodland is more or less continuous from here to where the river reaches its tidal limit 5 miles downstream and, as the river swings first left and then right, an imposing viaduct soars overhead carrying the main east coast railway line and giving the local name for this fabled stretch of river.

The river now meanders again beneath the steep banks of nearby Climbing Tree Farm, before settling back into a due easterly direction within the constraints of a tight, deep valley. Not far downstream of here, the footpath that has been following within a few yards of the right-hand bank begins to climb away up the valley side, making access to the river on this side far less straightforward. This problem can be solved by either wading across to take the footpath that has joined the riverside on the left bank or, if you'd prefer, by doubling back to the bridge at the entrance to Quarry Woods, crossing over, turning right and following the main road up the hill until you reach a waymarked sign for Bothal. This right turn is the footpath that will drop down to reach the far bank just before the viaduct, by which time the fishing is from either side. The long trek or the wade across could be well worthwhile, however, as while the dense bankside vegetation makes even the simplest of casting far more difficult, like the River Derwent at Lintzford, it is undoubtedly the inaccessibility and secluded nature of this place that makes it a haven for so many of the river's better trout.

After about two miles of overgrown glides and pools, the river starts to deepen on approach to another large weir just upstream of the small village of Bothal. The riverside path on the right bank has by now climbed all the way to the top of the valley, so access back down to the river is via the winding road down Bothal Bank. On the left bank, the footpath follows the riverside all the way.

We are now about three miles downstream of Morpeth (as the river flows), so if you weren't planning on following the river all the way through the woods, parking is possible by the roadside in Bothal village, provided due care is taken not to obstruct any access. This is the place where young Jack Charlton used to partake in his illicit trout fishing forays using a worm and ball of string, of course, but assuming you intend to use less rudimentary tackle, both fly and worm fishing is well catered for in this area.

The river begins to meander again for its last two miles as a strictly freshwater stream, as it slowly widens on approach to the tidal limit at Sheepwash weir. The Wansbeck Angling Association's waters end here, close to the appropriately named Anglers Arms in the village of Guide Post, yet there is one more opportunity for the trout angler to enjoy some sport before the Wansbeck becomes a languid semi-tidal estuary held back by an amenities weir a further three miles downstream.

Just below Sheepwash weir (and the bridge carrying the main Newcastle to Ashington road, the A1068), on the left bank, is the start of Wansbeck Riverside Country Park, a mile-long expanse of parkland that follows the ever-widening river as it bisects the former pit villages of Ashington and Stakeford. Inexpensive day tickets to fish for trout by any legal method can be obtained from the caravan site warden, although it should be noted that the further downstream from the weir you go, the slower and wider the river becomes. Access by foot is by following the path that heads downstream immediately on the northern side of Sheepwash Bridge, or alternatively by car, following the A1068 up the hill towards Ashington before taking the first available right turn into Green Lane and then turning right again to head back down towards the caravan park. Parking is free.

So, for the last time, as far as trout fishing is concerned: directions. By car, Morpeth clearly is signed on all approaches. The car park at Scot's Gill is reached from the south by following the main road all the way through Morpeth towards Alnwick, before turning left onto the B6343, or from the north by taking the A192 and turning right at the same point, before you reach Morpeth town centre. The car parks near Gas House Lane are found by following the same roads to the northern end of Telford Bridge (you will cross it coming from the south) and heading towards Ashington on the A197. To get to Bothal, continue on this road out of Morpeth for about two miles and at the second roundabout outside of town, turn right. Follow this minor road for about a mile. To get to Wansbeck Riverside Country Park take the A1068, which leaves the A1, heading north, at the Seaton Burn Interchange (signed A19 – take the second exit at the roundabout). Heading south, the A1068 is clearly signed coming off the A1 at Alnwick.

Buses to Morpeth are frequent from all points of the compass, but getting to Bothal or Wansbeck Riverside Park is less straightforward. The bus services from Newcastle Haymarket Bus Station to Morpeth are the 501, 505 and 518, while the X18 goes to both Morpeth and Wansbeck Estate, Stakeford, which is close to Wansbeck Riverside Park. All services are operated by Arriva. There is also an infrequent train service from Newcastle Central Station that calls at Morpeth and Pegswood (one mile from Bothal).

Wansbeck Angling Association doesn't currently have a website but reasonably priced permits (and day tickets) can be easily obtained from the local tackle shop, Game Angling Supplies, 3 Fawcett's Yard, Morpeth, NE61 1BG (opposite the new bus station), telephone 01670 510996, and also from McDermott's in Ashington (see NAF details).

CHAPTER THREE

The Return of the King: The Fall and Rise of Salmon Fishing in the North East

The king of fish in any river system is unquestionably the Atlantic salmon, and in the far North East of England we are fortunate to possess two of the greatest salmon rivers in the whole of Europe. The Tyne and the Tweed are each steeped in the traditions of this most prized of fish, both rich in a history of angling as well as the commercial netsman, but their respective stories down the last two centuries could not have been more different. The imbalance of blight and prosperity has at times been cruel and though divergent in fortune since the halcyon days of empire, the catch returns on these two great streams are at last starting to come together. With the industrial Tyne recovering and the pastoral Tweed continuing to flourish, once again, the outlook seems bright for salmon angling in Northumbria.

Throughout its hundred-mile expanse, the River Tweed passes through many important settlements but, unlike its southerly neighbour, it has always managed to remain largely unspoilt by pollution on its passage from the Central Southern Uplands to the North Sea at Berwick. Catch returns of migratory fish have never diminished, and thus, by the early twentieth century the Scottish border river was probably the most famous salmon fishery in the world, and with such a fine reputation the fishing became characterised by an exclusivity maintained on many beats to this day.

And the Tweed, a Scottish river for more than 50 per cent of its journey to the sea, and in England for the rest, is different – governed by its own unique set of byelaws drawn up by the River Tweed Commissioners, which also apply to tributaries like the Till, a river which flows exclusively through England. Even its main tributary, the Teviot, became a world-famous salmon stream in its own right and the confluence with the main river, at Junction Pool in Kelso, is a hallowed venue for the pursuit of large salmon often only days up from the sea.

Robert Blakey, in his 1854 tome *Angling; Or, How to Angle and Where to Go*, described the Tweed as 'one of the noblest fishing streams in Europe'. At the time he wrote, the rapidly expanding railway network was only just starting

The River Tyne at Tyne Green.

to make access readily available, but the famous river's reputation had already impelled at least one of its late-Georgian devotees to spring to verse:

> Along the silver banks of Tweed,
> 'Tis blithe the mimic fly to lead,
> When to the hook the salmon springs,
> And the line whistles through the rings,
> The boiling eddy sees him try,
> Then dashing from the current high,
> Till watchful eye and cautious hand,
> Have led his wasted strength on land.

The rhyme was simply attributed to *Glasgow, 1826* and, while it isn't clear whether the author was acquainted with William McGonagall, the reference does at least point to the importance of salmon fishing with rod, line and even fly in the years before Victoriana feted the Tweed with such sacred status.

Blakey considered the upper waters of Tweed 'a regular succession of fine streams and stretches of deep water, to which no pen can do anything like justice in the way of description'. However, he did caution that it wasn't until the reaches between Peebles and Kelso, where, 'The Tweed increases in bulk considerably; that here the salmon, and the salmon trout (sea trout), are to be met in much greater quantities than in the higher portions of the water.' He also described 'many splendid fishing stations where both salmon and trout can be readily captured with the fly', between Kelso and Berwick, and mentioned the Teviot, Whiteadder and Yarrow as rivers 'much frequented by North of England anglers, who find an abundance of sport in their waters during the whole fishing season'.

Blakey also described the River Till, the only Tweed tributary truly in England, as 'not a good fly-river ... a slow and languid running stream, very deep in certain localities, but containing very rich and fine trout'. It would appear, therefore, that by the 1850s, north Northumbrian folklore, passed down through generations of workers on the grounds of the Ford and Etal estates, had finally reached the ear of those who sought to advise the rich and honoured. The Tweed, they whispered, may be the best salmon river in the land, but the Till was only really a trout stream. Blakey's Victorian readership, travelling by train via Alnwick, thus continued north along the Millfield Plain until they reached the main river some ten miles further on and the locals kept the secrets of their Northumbrian 'trout stream' to themselves.

If, by the nineteenth century, the marketability of the Tweed as a salmon river had become as easy as the mere mention of its name, this hadn't always been the case. In his excellent book *Tyne Waters: A River and its Salmon*, Michael W. Marshall noted that Daniel Defoe had discovered, in 1725, that cured salmon sold in London and marked as 'Newcastle salmon' had, in fact, been caught on the Tweed and only brought by horse and cart to North Shields to be pickled. This was because it was the Tweed's more southerly companion, a river completely within English borders, which had the bigger reputation in those early days of union between England and Scotland. Two hundred and fifty years later, however, and the roles had been completely reversed.

In the foreword to Frank Johnson's *North East Angling Guide*, published in 1975, Charles Wade, angling correspondent to the *Sun* and secretary of the Northumbrian Anglers' Federation wrote, 'The mighty river Tweed is a mecca for salmon anglers the world over ... fishing can cost anything from £1 to £10 a day.' And this at a time when the full yearly salmon rod licence only cost £3.50!

The prices reflected the fact that, with the exception of one or two others in Scotland, the Tweed was one of the last rivers in the British Isles that still offered essentially the same opportunity to catch a big salmon as it had in centuries past. The Borders Regional Council's *Official Angling Guide* for 1979 summed it up like this:

To land a Tweed salmon can be the highlight of an angling lifetime. Whether it be in the spring sunshine amid the primroses and daffodils or on a chill October day with the fading leaves drifting slowly down, that day will remain indelibly etched on the memory.

They then got down to the hard sell:

The salmon fishing season on Tweed is lengthy, commencing on 1st February and ending on 30th November. The spring run is well established throughout the lower and middle reaches of the river on opening day and consists mainly of fish in the 6–12 lb class. Summer fish are slightly larger and in September, the famous Autumn run of large fish starts in earnest. From September to the closing date, the middle and upper reaches of the river normally hold large stocks of fish in the 18 to 25 lbs range, with occasional fish of much greater weight.

Such Tweed memories can be made of a single day or a single fish, but for certain lucky individuals they can consist of a whole career of close encounters. In his chapter for *Hooked on Scotland*, the companion book to the BBC TV series of the same name, Michael Shepley recalled his own varied memories from many years on the river.

My best ever salmon was a cock fish of 24lb, taken from Norham in October on a small jungle-cock tube fly fished on a sink tip line. I have hooked and lost similar sized fish, all of them on the Tweed. There was my 25-pounder (for that's my estimate) that followed my fly right to the bank on Upper Norham, and then swallowed it right at my feet before slowly swimming back into the depths. I lost him after 45 minutes. Then, a little further upstream on the Junction Pool at Kelso, I was wading the left bank at Ednam House. Gordon on the boat didn't see me hook the large cock salmon, which took off towards the bridge. Eventually the salmon tired and I brought him into the shallows. I tried to tail him where I was and had to tuck the rod under my arm to grasp the broad wrist with both hands. Halfway to the safety of the bank, my grip slipped and, at the same time, the fly fell out of the salmon's mouth. It rolled over twice and disappeared weakly back into the current.

Naturally, traditions hold true for a river on which the quality of the fishing has remained so good for so long. Sunday fishing is still not allowed on most beats (a custom common on most rivers in Scotland) and spinning is forbidden during the first two weeks of the season and from 15 September through to the back end. This rule corresponds to the period of the year when the salmon nets are in operation, but it has always been controversial, even back in the mid-seventies. In an article contained in Johnson's *North East Angling Guide*, Jim Hardy, of the Alnwick tackle firm, wrote:

On the Tweed, usually the water is fairly high at this time (the first fortnight of the season) and heavy tube flies are fished, which means that heavy lines are required to propel and turn the fly over correctly. This, in turn, results in the use of strong rods to carry the heavy line. This type of fly fishing is forced upon anglers as a necessity, rather than a pleasant way to fish. The water conditions are really more suitable for spinning, which is practised on the Coquet and Tyne from opening day ... so the angler is restricted to a method of fishing which bears absolutely no logic.

Yet despite concerns over the rationale of local byelaws and the occasional blip from year to year in salmon catch returns, the River Tweed has continued to go from strength to strength. In 2005, the Alba Game Fishing website, which promotes luxury fishing breaks in Scotland, proclaimed the Tweed 'the best salmon river in the world'. This was based on statistics which showed continued increases in rod catches on the river. They reported:

> The river Tweed is being hailed by many as the best salmon fishing river in the world, after a year in which a record breaking 15,257 fish were landed by anglers. The 2004 total was almost 10 per cent up on the previous year, according to the River Tweed Commissioners' (RTC) annual report. Netting stations on the river hauled in 2,991 salmon. Last year's haul topped 2003's record breaking 13,886 salmon caught by rod, which itself shocked even the most optimistic fishery manager. The 2003 haul prompted a number of angling experts to say the Tweed had become the best salmon river in the world.

Andrew Douglas-Home, the RTC chairman, echoed those comments: 'In these troubled environmental times, it is an extraordinary achievement. Seldom, if ever, has any UK river caught more salmon on rod and line.'

But the RTC report also warned of the thin divide between success and failure in the modern world of fisheries management, and Mr Douglas-Home said he had been reminded of the challenges faced at a recent seminar to discuss the threat posed by the deadly parasite *Gyrodactylus salaris* (GS). The Tweed's proprietors had been told that on the River Rauma in Norway, managers had been trying to eradicate 'this apocalyptic bug' for twenty-two years, in which time its salmon run had been decimated.

In 2008, the Scottish government outlined radical plans if it were ever required to deal with GS on the Tweed. Nick Yonge of the Tweed Foundation explained, 'If you are trying to get rid of it in a river system what you have to do is kill all the fish in the river – that is the only way of getting rid of it. GS can hook a ride on other species of fish – it doesn't affect them, it doesn't kill them or give them the disease.' He appealed to people travelling abroad to take extra precautions on their return and said it was vital to dry out any equipment which had used been in fresh water while overseas. He said it was important that attention was paid to all equipment, such as canoes, fishing rods or reels.

Mr Yonge was adamant that there is currently no alternative to the complete cull plans detailed by the Scottish government to stop the parasite. A special poison would be released to kill the parasites and their hosts and it is hoped parasite-free fish would than repopulate affected rivers. The plan would only ever be put into practice if an outbreak could not be contained, but it is a timely reminder of how such a pristine salmon fishery as the Tweed could be so easily destroyed.

* * *

In the mid-nineteenth century, at around the time the Tweed's reputation first started to go before it, the Tyne was equally regarded as one of the country's finest salmon rivers. Formed by the confluence of its North and South branches, the main Tyne is also a large river in the British sense, outsized by only the Thames and the Severn and equalled by few others besides the Tweed itself. What marks the Tyne out from all except its border neighbour is the fact that it, too, is a spate river fed only by other spate rivers throughout its course. As such, the Tyne can be assumed to have had a prodigious run of migratory fish since the Ice Age and from medieval times to the industrial revolution's zenith, the salmon fishing industry on the river certainly thrived.

However, the seeds of monumental decline had already been sown by the time the Victorian gentry first started their own yearly pilgrimage to the swift waters of the nearby Tweed. Industry and the sprawling metropolis that would soon become Tyneside began to foul the river's lengthy tidal range with choking pollution and as decades passed the catches fell. By the 1930s commercial salmon netting on the river was finished, destroyed by complaints from London fish merchants – the principal customer – that Tyne salmon tasted of engine oil. By the 1950s the salmon run on the river was effectively over as the stinking, lifeless estuary became devoid of oxygen and impassable to migratory fish.

The river's diehard salmon anglers became a lone voice of protest in that era of ecological apathy but, eventually, a reorganisation of water resources administration brought about a workable solution and out of despair came the river's salvation. Even so, it wasn't until the 1980s that counter-pollution measures along the estuary, combined with an ambitious salmon hatchery programme to compensate for the construction of the Kielder Dam, helped to restore a healthy run of fish to the Tyne. By the late twentieth century, the previously festering waters of the lower Tyne were clean enough that seals could follow the migrating fish as far as the tidal limit and, by the time of the millennium, the river and its north and south branches had all but recovered – once again regarded as the premier salmon river system to flow in England.

Even as early as the 1850s, the problems of early industrial pollution were beginning to show their hand. Blakey wrote little about the Tyne, save for a few sentences on the North Tyne, which he regarded as a 'first rate water', but he did report: 'The southern branch is nearly denuded of trout, from the effects of lead mines situated on its higher waters.' Nonetheless, the river system did, in fact, boast a prodigious run of salmon and sea trout at around that time and both anglers and the commercial netsmen recorded great catches in the years through until the early part of the twentieth century.

Indeed, the River Tyne was already possessed of several famous salmon anglers by 1850. One was Thomas Bewick, the famous wood engraver born in 1753 at Cherryburn, near the village of Mickley in Northumberland, a mere stone's throw from the south bank of the burbling main river. The Tyne here is still as beautiful as it must have been in the late eighteenth century, descending a series of wide riffles down to a half-mile-long deep, and bounded on the Cherryburn side since 1838 by the railway wall of the main Newcastle to Carlisle line. This beat, controlled until recently by Big Waters Angling Club, is a prodigious holding pool for migrating salmon and trout and it comes as little surprise to learn that the young Bewick found it almost impossible to resist its piscatorial charms.

In Bewick's autobiography, *Memoir*, published in 1862, after his death, he recalled his early angling expeditions that, even then, appeared to follow the traditional Northumbrian trout fishing season of March to September.

As soon as the bushes and trees began to put forth their buds and make the face of nature look gay – this was the signal for the angler to prepare his fishing tackle, and in doing this I was not behind hand with any of them in making my own all ready. Fishing rods, set gads and night lines were all soon made fit for use and with them late and early I had a busy time of it during the long summer months and until the frosts of autumn forbid me to proceed.

However, like generations of young anglers to follow, his exploits appear to have been the cause of great anxiety to his parents.

The uneasiness with which my late evening wadings by the waterside gave to my father and mother I have often since reflected on with regret – they could not go to bed with hopes of getting to sleep, while haunted by the apprehension of my being drowned, and well do I remember to this day, my father's well known whistle which called me home – he went to a little distance from the house, where nothing obstructed the sound, and whistled so loud through his finger and thumb that in the still hours of the evening, it may be heard echoing on the vale of the Tyne to a very great distance. This whistle I learned to imitate and answered it as well as I could and then posted home.

Bewick's name is synonymous with his two classic books, *The General History of Quadrupeds* and *History of British Birds*, but, according to a fascinating article by Keith Harwood in the angling magazine *Waterlog*, he was working on a third manuscript, *A History of British Fishes*, at the time of his death. Sadly, this book never saw the light of day.

By the time Bewick had entered his later years, his beloved river was already starting to show the first signs of its slow decline. In *Tyne Waters: A River And its Salmon*, Michael W. Marshall noted that, in 1807, the price of Tyne salmon had risen from its time-honoured rate of 1*d* a pound to as much as 3*s* 6*d*. This was because of a marked downturn in the catches of salmon from the many netting stations located on the lower river and Dr Marshall records that the decrease in numbers was attributed to a variety of reasons – a lock at Bywell that held up or even prevented the fish from ascending to their redds and an increase in the number of craft in the estuary. Most pointedly, Marshall quotes a Mr MacKenzie, in 1827, as having blamed 'the deleterious mixtures that are carried into the stream from the lead mines and various manufactories on the banks of the river'.

Later on in his book, Michael Marshall outlines the way dredging, which commenced in the mid-nineteenth century, had such an adverse effect on the natural cleansing action of the tide throughout the estuary. This came to coincide with the arrival of newly created sewer networks in the rapidly expanding towns along both riverbanks, and effluent could linger for up to ten days or, worse still, be forced upstream on incoming tides to settle on the riverbed. In addition to this, industry was discharging poisonous effluents right throughout the length of the river's tidal range. By the middle of the Victorian period, concern was already being shown and Marshall quotes Grimble, in 1887, as having predicted that 'the Tyne might become as salmon-less as the Bristol Avon, Mersey, Aire, Calder, Weaver, Taff and many smaller rivers from which the salmon has been exterminated solely by poisonous pollutions'.

Yet salmon continued to be caught in fair numbers on the Tyne, on both rod and line and in the nets, right through to the interwar years in first half of the twentieth century. There were good years and bad, and the better catch returns tended to coincide with greater than average rainfall and regular spates that diluted the pollutants in the tidal stretch. Scientists later discovered that fresh water forms a layer or 'wedge' over the top of brackish water through which migratory fish can travel. For the time being, this was enough to get some salmon through, but increasing abstraction to supply water to the new centres of population (which had included, in 1900, the construction of Catcleugh Reservoir in the headwaters of the River Rede) was slowly emasculating the once mighty main river and reducing spates. Catches peaked at over 3,000 rod-caught migratory fish on the Tyne in 1927, but the inevitable demise of the river's ancient salmon fishery was soon to be at hand.

Salmon catches on the Tyne finally bottomed out during the 1950s. Environment Agency figures show very few rod-caught salmon recorded during the decade, a situation which continued on until the mid-1960s. With the last river nets having gone out of operation in the 1930s, the statistics suggested that vanishingly few adult salmon were making it through the polluted estuary, while the progeny of those that did succeed in spawning faced similar problems on their arduous journey to the sea. The River Tyne was on its last legs as a salmon fishery and only drastic measures were going to save it from oblivion.

Salvation finally came with the formation of the Northumbrian Water Authority in the early 1970s, one of ten such bodies in England and Wales whose role it was to oversee all aspects of water management – in their service divisions, water supply and sewerage, and in their environmental capacities, all aspects of water quality policing and fisheries. Many instantly saw a potential conflict of interests but, as far as the Tyne salmon and her anglers were concerned, Northumbrian Water Authority's eighteen years of existence marked the rebirth of their great river.

NWA's solution to the pollution in the estuary was to build a system whereby sewers that had previously discharged directly into the Tyne would instead flow into two interceptor pipes laid on either side of the river. The interceptors on the south bank would run west from the coast and east from Ryton, meeting at Scotswood Bridge under which their combined load would be carried over the river. On the northern side, the pipe crossing the bridge would discharge into the interceptor running parallel and the waste would then be carried a further 10 miles east to a treatment plant at Howdon.

The scheme came into full operation in the late 1980s, completely freeing the Tyne of sewage pollution. The accumulated effluent now discharges into settlement tanks in which micro-organisms act on organic pollutants, reducing the biological activity of the waste and producing a sludge which can be discharged onto ships and dumped at sea around 10 miles offshore. The process is self-perpetuating, as some of the by-product retains biological activity ('activated sludge') and can be reused to treat the continuous accumulation.

The benefits of the new system were manifestly apparent. Annual rod catch returns, which had recovered to the tune of 200 to 300 from the late 1970s to early 1980s, consistently broke 500 as the system came on line in the latter part of the 1980s. By the 1990s, with the scheme fully operational, there were reported catches exceeding 1,000 every year, bar one, and the turn of the millennium heralded consistent figures of over 2,000 every season. In 2003, 2004 and 2005, annual catch returns broke 3,500 and the Tyne was in its best form since records began in 1880.

Yet the Water Authority's lasting legacy is not limited solely to the Tyne Sewage Treatment Scheme. Just as the first pipes were being laid down on

Tyneside, bulldozers moved into the pastoral North Tyne valley between Bellingham and Kielder and began construction of what was to become, in 1982, the largest man-made lake in Western Europe.

Apart from two reservoirs, Kielder and Bakethin (which are one and the same in times of high water), this project also encompassed the Kielder Water Transfer Scheme whereby water stored in the new lake could be transferred as either supply water or 'compensation flow' to the rivers Derwent, Wear and Tees, whose headwaters were all already heavily abstracted. While many had reservations about such a system, compensation flow was reported to be of potential benefit to game anglers on all these rivers, as it was intended only to release it in times of drought. Runs of summer salmon could thus be 'kept on the move' with regular releases, whereas before they would be stuck in the same pool for weeks, losing strength, condition and angling potential.

However, there was also now the problem of the 170-foot-high dam thrown across the North Tyne just upstream of Falstone. The installation of a fish pass was considered too expensive and also impractical, being so high up such a major river system. This excluded all migratory fish from the upper part of the North Tyne and, therefore, all the principal spawning grounds of its newly rejuvenated salmon run had been effectively lost.

The solution was a hatchery at Kielder, where salmon parr could be reared from the fertilised eggs of adult salmon captured in the North Tyne. The operation was soon electro-netting salmon below the main dam, stripping them of their eggs or milt and releasing them unharmed to commence their journey back to the sea. By 1989, according to Michael W. Marshall, 320,000 salmon parr were being released into the Tyne system each year, undoubtedly a major contributory factor to those ever-improving rod catch statistics.

Countless salmon anglers have now reaped the benefits of this ingenious scheme, from the rich and the famous to just plain ordinary local fishermen. One such person was Tony Gleeson, who back in the mid-1990s enjoyed one golden summer under the spell of the Tyne's silver tourists.

In 1996, Tony had relocated from Yorkshire to the Northumbrian village of Acomb, to take up the position of charge-nurse at Hexham General Hospital and he wasted no time in adding his name to the waiting list for the local fishing club. Acomb Angling Club held the north bank of the River North Tyne, which also became the north bank of the main Tyne where it joined the other principal tributary at a place known locally as 'The Meetings' – the River Tyne's equivalent of Junction Pool. As luck would have it there was a vacancy, Tony became a member, and for him the summer of 1996 was about to become one he would never forget.

The first Saturday of August saw Tony at a loose end, so he'd decided to spend the day fishing down at The Meetings. With it now well into the second month of the coarse season, he spent most of the first five hours bagging up

A fish ladder on the River Tyne designed to assist the passage of salmon and sea trout.

with dace, roach and the odd small trout, using light stick-float tackle in the steamy water on his near bank – this kind of fishing was, after all, his bread and butter. But, after a few hours, something strange started happening. Tony's expertly guided float tackle, which he'd been trotting through the 4-foot-deep channel without a hint of trouble for four hours, suddenly started veering left into the near bank and snagging. There was nothing he could do to prevent it and it was ruining what had, up to then, been a fantastic session.

He was just considering whether to switch to using a heavier avon float, when there was a giant splash near the tail of the pool. A few moments later came another crash of big water; this time accompanied by an unmistakable flash of silver in the sunlight ... salmon! Tony abandoned his coarse fishing and immediately got out his 11-foot, 1½-pound test curve 'whopper-stopper', crimping a couple of AAA shot just up the line from the large devon to give him the extra casting weight required to cover the wide pool.

By the time Tony was ready to start, the reason for his earlier problems with the float gear was now clearly apparent. In the pool below The Meetings, usually dominated by the racing current from the North Tyne, the roles had been unmistakably reversed. The South Tyne, whose headwaters were still those of a natural wild spate river, was on the rise. The storms that had passed north over the border in the early hours of the morning had deposited several inches of rain on the fells and this had now flowed the length of the South Tyne and on into the main river.

A rising river was a killer to coarse fishing but, in any case, Tony was now setting about the task of trying to provoke a take from that salmon that had made all the noise about half an hour earlier ... if only it was still in the pool. The approaching spate had obviously roused it from its slumber and he just hoped it hadn't already taken a left turn and headed up the South Tyne in search of its redds.

Cast after cast he made, methodically trying to cover all the water in the deepest part of the pool. On the third cast of his second spell, there was an arm-wrenching snatch and Tony's powerful fixed spool reel started screaming, the 12-lb line tearing off the spool as a large salmon bored head down for the tail of the pool, seemingly with all the resolve to just keep on going until it got past the Tyne Bridge!

After what seemed like an eternity, he managed to turn the fish and it came back towards him, breaking surface and flexing in mid-air, before crashing back into the water about five yards away with the sort of splash that had betrayed its presence in the first place. Still in contact!

Eventually, the fish tired and Tony managed to get it into the shallows on the near bank, where he hoped it was less likely to make one last dash for freedom. Still, the salmon thrashed its powerful tail at the short line while, somehow, its captor managed to throw his landing net into the margins and then thrust it under the fish. She was his, 18½ lb of pure silver, not quite straight from the sea, being so far upriver – but fresh run all the same.

A week later, with the fading effects of the spate still keeping the Tyne salmon on the move, Tony caught another, same place, same method – only a bit smaller this time at 13 lb. He didn't catch any of the more revered autumn salmon that year – fish that can run well into the twenties and more – but he wasn't bothered. Some folk might have expected more for £120 but for someone more used to 'slumming it' on muddy beaches and coarse fishing rivers, a season marked by two double-figure salmon was well worth the money!

* * *

This was a defining moment. What they'd once said about a Tweed salmon marking the highlight of an angling lifetime had certainly held true for Tony Gleeson on the Tyne in 1996. He only lived in the area for a couple more years, but those two fish would characterise the entire stay. For Tony, it was soon to be back to the 'day job' of coarse and sea fishing, but the old Tyne had shown the extent of its newfound potential and stood on the brink of a whole new era.

Other rivers in the area can tell a similar tale and the Wear has made an almost identical recovery from pollution to that of the Tyne, seeing not just the return of its salmon run, but also elevation to the status of one of the best sea trout fishing rivers in all of England. Smaller streams can also boast the return of this migratory relative of the salmon, and down in the far south of the region, the Yorkshire Esk, which runs off the North York Moors down to the sea at Whitby, boasts prodigious runs of both sea trout and salmon. Nowadays the Esk is the county's only salmon river, having been held in high regard since 1894 when Tom Bradley proclaimed, 'The whole of the fishing is high class – salmon, sea trout and brown trout,' in the second edition of his *Yorkshire Anglers Guide*. One hundred years later, it is still being feted.

Back up in Northumberland, the River Coquet also bucks the trend, having, like Tweed and Esk, never lost its ever-reliable run of migratory fish. In the years when the larger rivers had given up the ghost as far as salmon and sea trout fishing were concerned, this delightful stream continued to provide affordable fishing in spectacular surroundings, right on the doorstep of the North East game fisherman.

Robert Blakey thought the Coquet 'one of two or three streams in Northumberland of first rate angling note'. He continued:

> The Coquet is the most celebrated, and has for more than a century been a stream enjoying aristocratic and fashionable notoriety as an angling locality. In former years, before the fashion ran so strongly for distant Scottish rivers, the Coquet used to be the annual rendezvous of all our London literary, scientific and political fishers; and even now there are more anglers on its streams, and more fish taken out of them, including the salmon trout, than in any other half a dozen of chief rivers in the Northern counties of England.

The river's run of migratory fish was, as it still is, a major attraction: 'There are no artificial or natural obstructions for the free passage of the fish from the sea to its highest waters, so this noble fish (the salmon) can always be found, in more or less abundance, in every section of its waters.'

But above all, Blakey thought the Coquet's greatest charm its 'range of almost 40 miles, all of which is open water for the angler, with the exception of three of four small sections of it; and these, even, are not very rigidly preserved. This freedom from constraint of every kind is a pleasurable element in piscatory recreations.'

He also observed that, 'The salmon fishery at the mouth of the river belongs to the Duke of Northumberland, and is let for considerable rent; but we have never known any angler called into account for capturing the salmon with rod and line, wherever he might be perambulating on the Coquet.'

In 1975, Frank Johnson described the Coquet as 'the North east's premier salmon river, surpassed in the region only by the famous Tweed'. That this is no longer the case is in no way a poor reflection on a river that now averages yearly rod catches almost five times that of the mid-1970s. Rather this change in status has been brought about by the huge catches recorded on the Tyne.

Logged rod-caught salmon returns of more than 4,000 for the Tyne in 2004, while still bearing little comparison with those for the Tweed, did represent a monumental improvement over the nil catches for the river recorded from 1955 to 1961. The Environment Agency bar chart for declared salmon returns between 1952 and 2007 shows the situation clearly, the graph scraping along the x-axis for fifteen years before gradually improving in the 1970s and then soaring into the thousands from the late 1980s onwards.

Peaking at over 1,000 salmon in 2004, the profile for similar statistics (1965–2007) for the River Wear is more or less the same, while returns for the Coquet (1952–2007) show a far greater level of consistency. Averaging 500 in the 1950s and 1960s, and from the mid-1980s on, the declared yearly rod catches of between 100 and 400 for the years between 1968 and 1986 probably reflected factors such as commercial netting along the North East coast. Attrition rates were high in those years, but fortunately the harvest of returning fish from the seas has been greatly reduced.

The commercial North East Coastal Salmon Fishery, comprising mainly drift nets operating up to 6 miles offshore, used to be the single biggest net fishery in England and Wales. The introduction of monofilament nets to the industry in the 1960s culminated in a declared catch of 90,000 salmon and sea trout in 1970, with an average of around 50,000 per year taken from the late 1960s to the mid-1980s. Not surprisingly, this had to have a major effect on the numbers of fish returning to spawn and prompted more draconian fisheries management.

In 1992 and 2002, Net Limitation Orders were introduced and in 2003 a permanent buyout of most of the drift net licences held for the North East coast was secured. The number of licences issued by the Environment Agency fell from 142 in 1992 to 16 by the time of the buyout, corresponding with a reduction in the EA's reported average yearly catch returns from around 30,000 in the early-1990s to a mean of about 7,000 from 2003 to 2007. The Agency has estimated that this represents a potential saving of 22,000 salmon that might otherwise have been prevented from re-entering the North East rivers.

And thus it can be seen how the health of the salmon population in the major rivers of Northumbria continues to improve. Through a combination of water quality management, hatchery rearing and the rationalisation of the coastal fishing industry, the main problems to have blighted the North East's salmon rivers over the last 200 years have been circumvented. Newer issues, such as GS remain but, for the time being at least, all is well on the rivers Tweed, Tyne, Coquet, Wear and Esk. Even the Tees and the main Yorkshire rivers have seen slight improvements in their status as 'former' salmon fisheries, indicating the first signs of success in the latest round of estuarine water-quality initiatives. Things are looking up and there was certainly never a better time to be a salmon angler in the North East of England.

CHAPTER FOUR

Beat Fantastic: A Brief Guide to Salmon Fishing in Northumbria

By now of course, like with the trout fishing, you should be straining at the leash – your appetite well and truly whetted – and you'll be keen to find out exactly where to go to have the best chance of pitting your wits against the undisputed king of the river. Over the years, salmon fishing has tended to become more and more of a holiday pastime, its expense and exclusivity often taking the best fisheries beyond the financial scope of a mere club or association. While there are some club beats where the angler fends for his or herself (more particularly on the North and South branches of the Tyne, the Coquet and the Wear) many of the best salmon fisheries offer guides and/or ghillies to help the eager fisherman get the best of what's on offer.

Many salmon anglers therefore like to book one or more of what are usually a strictly limited number of places to fish for a day or a week on just such a commercial beat. The fishing is usually sold as an all-inclusive package – the price dependent on a particular fishery's reputation – with all kinds of extras, from food and accommodation to ghillies, either included in the price or offered at extra cost. It is rarely inexpensive.

Therefore, rather than the sort of guide provided for trout fishing, there now follows a brief overview of some of the more celebrated salmon beats on the rivers Tyne, Tweed and one or two of their tributaries. Contact details for making reservations are included, although it should be noted that not all the beats have places available all of the time.

For starters we have one of the North East's finest salmon beats, a stretch of river that used to be controlled a by one of the region's most highly regarded angling clubs. Up until the late 1970s, the River Tyne's famous Bywell Syndicate was leased by none other than the Northumbrian Anglers' Federation, but long before the recent surge in reported rod catches on the river, the Federation had lost control of Bywell and now local landowner Allendale Estates operates the beat.

Bywell consists of around 2½ miles of the main River Tyne between Prudhoe and Corbridge, which can be accessed from either the nearby A69 trunk road or the A695, which runs to the south of both the major road and the river. There are twelve named pools on the river here, some of which are long, swift and powerful, as well as several slower, deeper stretches that fish better in periods of high water. The whole stretch is fished on a two-beat system, with four rods rotating daily. There are two fishing huts, two ghillies and two boats – with the recent introduction of boats having reportedly opened up hitherto un-fishable water. The five-year average for salmon and grilse on the beat, up to 2008, is 350 and Allendale Estates offer excellent self-catering accommodation within walking distance of the river.

To check availability and reserve fishing at Bywell, phone 01661 843296, write to Bywell Home Farm, Stocksfield, Northumberland, NE43 7AQ, or email lindablair@allendale-estates.co.uk.

The main Tyne at Bywell is, of course, influenced to a large extent by water releases from Kielder, but upstream on the North Tyne this is even more of a factor. Another well-known Tyne salmon beat is Chipchase, situated on the North Tyne between Wark and Barrasford, with a five-year average of sixty-six salmon and grilse and an almost equal figure for sea trout. The sea trout run on Chipchase is predominantly in June and July, and while salmon are encountered from April onwards, their capture predominates towards the latter end of the season.

Chipchase consists of three beats and is best suited to fly fishing, the 'upper' and 'lower' beats each consisting of seven named pools. At the upstream end, the pools are Comogen, Crow Pool, Causeway, Top Straights, Bottom Straights and the Top and Tail of Catherine's Pool. Below this are the highly regarded Mill Stream, Little Island, the high-water pool Martigan, Strother Croy, Nunwick Mill Stones, Ash Tree and the Bottom Salmon Pool.

The third beat at Chipchase is called Riverhill, which averages a slightly lower five-year mean than its 'upper' and 'lower' counterparts. There are eight more pools here, equally well suited to the fly, beginning with the slower Pete's Rock and continuing through Scar Pool, Rack, Coldwell, Cindy Wells, Stepping Stones, Harry's Wood and Ellwood Pool.

Access to Chipchase is via the B6320 Bellingham road, which leaves the well-signed B6318 'military road' where it crosses the North Tyne at Chollerford. For full information on availability and bookings, contact Smiths Gore Estate Agents, Eastfield House, Main Street, Corbridge, Northumberland, NE45 5LD or phone 01434 632001. There is a fishing hut, and a ghillie may be available by arrangement.

Just downstream of Chipchase, also on the North Tyne, is Chesters, a fishery of considerable renown that takes its name from the nearby ancient monument, Chesters Roman fort. This beat is situated downstream of the B6318 road bridge at Chollerford, is strictly fly-only except during high water, and divided into two separate beats. There are a maximum of two rods allowed per beat,

Haydon Bridge on the River South Tyne.

with fishing on all days except Sundays and there are fishing huts, a boat and ghillies available. Information can be obtained by contacting Aidan Pollard on 07753 729192 or emailing aidanpollard@hotmail.co.uk. Aidan is the former manager of the Bywell Syndicate and now runs the excellent Reeltime Fishing and Leisure Services, more about which can be found on their website: www.reeltimefishingandleisure.co.uk/

Needless to say, the South Tyne is also excellent for salmon fishing, although unlike the north and main rivers, it is a natural spate river, whose runs of fish are influenced solely by season and rainfall, and not by reservoir compensation releases. The best known fishery on the South Tyne is the Lambley Estate, situated almost 30 miles upstream of its confluence with the North Tyne, and accessed by taking either the A689 from Alston towards Brampton or by leaving the A69 on the Haltwhistle bypass and heading towards Lambley.

Lambley comprises about three quarters of a mile of double-bank fishing and, when available, can be fished by up to four rods at any one time, with four named pools: Birkett's, Wallace, Madam's and Jackie's. This section of the South Tyne runs mainly in a south–north orientation and the current average rod catch is ninety-eight fish (salmon and sea trout) per season, in spite of the fact that the beat is not that heavily fished. The best of the sport usually occurs from late May onwards, with any prolonged spell of rainfall bringing the fish up. More information is available by phoning 01434 322121 or by emailing tynesalmonfishing@live.co.uk.

Of course, not all of the salmon fishing on the three main rivers in the Tyne catchment is restricted to syndicate- or estate-managed beats. So, very briefly, here is a list of some waters controlled by local authorities or angling clubs – membership of which may or may not be subject to a waiting list:

The Wylam Angling Club controls fishing at Wylam and Hagg Bank – information and day tickets are available from the Spar shop in Wylam village or by phoning 01661 852214.

The Northumbrian Anglers' Federation controls two sections of the main River Tyne near Wylam and Ovington – contact details are the same as those for trout fishing on the River Coquet.

The local council at Hexham sell day tickets for the mile-long Tyne Green on the edge of the town, comprising both fast runs and deep water less than a mile below the meetings of the North and South Tyne. Visit the Tourist Information centre in Hexham or phone 01434 652220 for more details.

Lastly, the Haltwhistle Angling Club have 6 miles of the River South Tyne available for salmon fishing at £20 per day, £50 a week or £120 for a season ticket. Permits are available from the club website, www.haltwhistleangling.co.uk.

* * *

Moving north, we already know that the River Tweed has a longstanding worldwide reputation for both the quality and extent of its salmon fishing. And with a total length of 98 unspoilt miles from source to sea and a catchment of some 1,500 square miles, is it any wonder? With its nationwide reputation for sport fishing having lasted uninterrupted since the seventeenth century, it is no surprise that the banks of the Tweed comprise some of the most exclusive fishing rights on the planet. Prices can be astronomic – this river's beats are to angling what Gleneagles and the Royal & Ancient are to golf! Let's have a look at some of what's on offer, either on or not too far away from the lower 'half-English' waters of this mighty stream.

The first Tweed fishery that has to appear in any guide is the most famous one of all – Junction Pool. This marks the confluence, at Kelso in the Scottish Borders, of the Tweed and its almost equally famous tributary, the Teviot, with both rivers already forming mighty flows long before their meeting point. The Junction fishery comprises 1½ miles of some of the best double-bank salmon fishing in the world and apart from Junction Pool itself, there are ten other named pools: The Hawthorns, The Pot, The Flats, Bridge Pool, White Dyke, The Stream, Garden Foot, New Bridge, Rose Bank and The Grain. In addition to the fishing on the Tweed, the Junction beat also includes 350 yards of bank at the very bottom of the Teviot, featuring Jack's Plumb Pool.

The Junction fishery boasts a staggering five-year average of 817 salmon and grilse, there is a fishing hut (situated beside the Teviot) and boatmen and ghillies are available in the spring and autumn, which corresponds to fly only restrictions. The beat allows four to six rods, although its extreme popularity means it is usually heavily booked in advance.

Kelso is situated about ten miles on the Scottish side of the border. Access from the English side is by the A698, which is found by following the A697 (which leaves the A1 just north of Morpeth) over the border, and on through Coldstream. The A698 to Kelso branches off to the left about a mile beyond Coldstream. Alternatively, take the A696 and then the A68 from Newcastle to Jedburgh, continue on for about five miles then turn right onto the A698. This brings you into Kelso in the opposite direction. The contact for bookings is: Strutt & Parker Estate Agents, 55 Northbrook Street, Newbury, Berkshire, RG14 1AN, telephone 01635 576905 or email markmerison@struttandparker.com.

Carham is approximately 9 miles downstream of Kelso and begins where the border between England and Scotland descends from the Cheviot Hills to follow the River Tweed for all but the last 4 miles of its course to the sea. Carham is the first settlement on the right (south) bank of the Tweed that is in England, although the fishery comprises both banks, on the English and Scottish sides.

There are thirteen named pools on this mile-long beat, with the river twisting and turning on its approach to Coldstream. The five-year average for Carham is 298 salmon, with catches reported throughout the season, including during periods of low water. In these conditions, the whole beat is well suited to wading, although two boatmen are provided throughout the early and late periods where the fly-only rule is in force. There is also a fishing hut.

Carham is found by following the B6350 from either Kelso or Cornhill-on-Tweed (a mile from Coldstream at the English end of the A697) and turning in at a white gate beside the church. Further details are available by telephone on 01890 830200, or email peter@carham.net.

Not far downstream of Carham on the English side is South Wark fishery, consisting of about 2½ miles on the right bank only, with a five-year average of 330 salmon and grilse. There are seven named pools on this beat with accommodation for three rods, with two boatmen, who also double as ghillies. The fishing hut comprises an old boathouse overlooking the river in the shadow of Wark Castle and the whole beat is best fished in medium to high water. Access is from the B6350 about 2 miles east of Carham. The contact address is Wark Farm, Cornhill-on-Tweed, TD12 4RE, or telephone 01890 882862 for more information.

The next beat downstream on the English side is West Learmouth, which is administered by Fishpal and situated about a mile upstream of Cornhill-on-Tweed. According to its webpage, Learmouth has been one of the star performers on the Tweed in recent seasons, consisting of about two thirds of a mile of the right bank with a five-year average of 320 salmon and grilse. Success stories from West Learmouth include one angler who took twelve salmon in a single day and two others who shared fifty-two fish in a five-day session, reflecting the reason why this beat is reputed to account for more salmon per yard of bank than any other part of the Tweed.

The beat takes two rods, with a ghillie, boat and a fully equipped fishing hut provided and fishes best in conditions of medium and low water. Prices range from £60 to £750 per rod day and the contact details are: FishPal, Stichill House, Kelso, TD5 7TB, telephone 01573 470612 or email: info@fishpal.com. The webpage for West Learmouth is www.leeming.co.uk/pages/westlearmouth.asp.

From Cornhill, it is only another 5 miles downstream to the mouth of the River Till; the Till is the only major Tweed tributary whose catchment is confined entirely to England. Tillmouth itself provides one of the main river's finest salmon fisheries with 4¼ miles of the right bank and an average of 771 salmon and grilse caught between the years 2004 and 2008. According to the leasing agents for this stretch, Sale & Partners of Wooler, 'The Tillmouth fishings extend for 4.14 miles of single English bank downstream from Coldstream Bridge and have 22 named pools which hold fish throughout the spring, summer and autumn seasons. The beat extends through an area of outstanding natural beauty and is sheltered by a number of spectacular river cliffs and wooded banks.'

Being only fifteen miles up from the sea, Tillmouth has very large runs of fresh salmon throughout the season, with easy wading from accessible banks complemented by a fleet of twelve boats spread between most of the main pools, with six full-time boatmen employed. There is a fishing hut and a maximum of six rods are permitted, with priority booking given to parties taking a full week's fishing – a decision that could prove fruitful, as in one week during the record 2007 season (1,052 salmon caught) an unbelievable 102 salmon were taken.

Tillmouth is located on the A698 between Cornhill and Berwick. Contact details are: Sale & Partners, 18–20 Glendale Road, Wooler, Northumberland, NE71 6DW, telephone 01668 280803; or alternatively, contact Strutt & Parker, at the same address as for the Junction beat at Kelso.

If you were to take a short detour up the Till itself, you would find several good fisheries in its lower reaches. The first place you come to is Tiptoe, located just 2 miles upstream from the confluence with the Tweed, a beat whose five-year average is thirty salmon and grilse. Similar to but smaller than the north and south branches of the River Tyne, the Till is also a very good sea trout river, with a corresponding five-year average of twenty-one at Tiptoe for the salmon's smaller cousin.

Tiptoe has fourteen named pools and is a mixture of streamy and slower water extending for a mile on the right bank, the wooded bankside terrain described as being best suited to the 'able bodied' angler. There are a maximum of two rods including day lets and there is a fishing hut positioned midway down the beat with a bench and a shelter. Access is from Tiptoe Farm, which is reached by turning off the B6354 Berwick to Millfield road in Duddo. The river is then a 200-yard walk down a fairly steep track. The contact address for this fishery is Old Egypt, Tiptoe, Cornhill on Tweed, Northumberland, TD12 4XD, telephone 01890 883060.

Another well-regarded fishery on the lower Till is the Tindall beat, which comprises two sections, upper and lower, which are each fished from opposite banks. The upper Tindall beat, immediately downstream of the small village of Etal, is fished from the left bank and has sixteen pools. The Till at this point is faster-running than on most other sections, with swift rapids feeding some deep pools and making for ideal fly fishing conditions, although worm and spinner are allowed. Access to the car park for Upper Tindall is down a woodland track that leaves the B6345 just north of Etal, with a bridge providing access to the left bank.

Lower Tindall begins on the right bank immediately downstream of the upper section. Access is from a car park at Tindall House Farm, which is found by following the same side road off the B6354 as for Tiptoe, but turning sharp left after about a mile at a crossroads and then taking the first right turn towards the farm. This beat is 2 miles long and, running through a wooded gorge, the river is similar in nature to Upper Tindall, with twenty pools. The Tindall beats have a five-year average of seventy-four salmon, grilse and sea trout and prices range from £20 to £25 per rod per day. Information about Upper and Lower Tindall is available from Redscar Cottage, Milfield, Wooler, Northumberland, NE71 6JQ, telephone 01668 216223.

Back on the main River Tweed, downstream of Tillmouth we are now very much into what is termed the 'bottom' section of the river. There are several more highly regarded salmon fishing beats on both sides of the river before the tidal stretch is reached at Berwick and the last few salmon netting stations still in operation on the Tweed are encountered.

About 5 miles below the Till, near the village of Norham, the Pedwell beat has 1½ miles of prime salmon fishing on the English bank, bordered at its downstream end by Norham Bridge. Pedwell has a five-year average of 104 salmon and grilse and consists of seven named pools that can be fished by up to four rods, with a reputation for excellent fishing in even the most extreme low-water conditions. Pedwell can be fished on Mondays, Tuesdays, Fridays and Saturdays throughout the season and there is a ghillie to assist the anglers.

The River South Tyne.

It is accessed by taking the B6470, which branches off the A698 near Norham. Booking, as for Tillmouth, is by contacting Sale & Partners in Wooler.

Another 5 miles and several wide sweeping bends downstream of Norham, the tidal limit of the Tweed is reached at Union Bridge, just below the village of Horncliffe. The Horncliffe salmon fishing beat has 2½ miles of the right bank upstream from Union Bridge and averages ninety-nine salmon and grilse for the five years up to 2008. The beat runs from Norham Castle to the mouth of Horncliffe Mill Burn, half a mile upstream of Horncliffe. Like Pedwell, it fishes best in low water and sees its lowest pools, below St Thomas' Island, influenced by very high spring tides. Horncliffe accommodates six to eight rods and has a ghillie and a fishing hut. Contact details for this beat are from Fishpal at the same address as for West Learmouth.

Horncliffe is the last angling station on the English side. About 8 miles downstream of Union Bridge, the Tweed enters the sea and there ends a hundred miles of arguably the greatest salmon fishing in the world – hope you enjoyed the journey!

CHAPTER FIVE

Reservoir Dogs: The Rise of Stillwater Trout Fishing in the North East

If four consecutive chapters about fishing on rivers has got the sound of running water permanently ingrained on your hearing, now is the time to imagine instead a stiff breeze and crashing waves onto a rocky shoreline, or that eerie stillness that comes all of a sudden on a midsummer evening when the sounds of the countryside carry over what can seem like miles of glassy, flat, calm water. For what exploring a river in pursuit of trout or salmon gives in the way of its own indescribable attraction, the more modern branch of game fishing for stocked rainbow trout on reservoirs and other large bodies of water can equal with a quite different sort of appeal. Stillwater trout fishing is hardly new; for several centuries it has held a place in the fine traditions of game angling. However, whereas in years gone by it was confined to those parts of the country that had large natural lakes, like Scotland and the Lake District, there are now many more opportunities for the angler thanks to the steady growth throughout the twentieth century in the numbers of reservoirs built for water supply.

A reservoir is defined as a large natural or manmade lake for the collection and storage of water ultimately to be used for domestic water supply. In the modern water supply industry, there are two distinct kinds of reservoir, each designed for a specific purpose with regard to the containment of groundwater. Direct-supply reservoirs store water that is to be transferred straight to a water treatment works, their contents having usually been piped or pumped in from elsewhere. River-regulating reservoirs, meanwhile, impound water collected from rainfall – usually by means of a high dam – so it can be released, as required, to compensate for water removed further down the dammed-off river for supply. These releases are known as compensation water (or flow), usually drawn from a valve situated in the deepest point of the reservoir and released into the headwaters of the re-emergent river at the foot of the dam.

River-regulating reservoirs are found in parts of the UK where the relief of the land is irregular, principally at the headwaters of spate rivers, which have a far greater catchment owing to the nature of their surroundings.

The vast acreage of Kielder Reservoir.

Their construction allows water companies greater freedom to remove water from the lower reaches of such rivers (a process known as abstraction), as compensation water can be relied upon to keep flow rate at a fairly natural level for most times of the year.

It therefore follows that in the North East there are a greater than average number of such water features, ranging from the very large and deep river-regulating reservoirs at Kielder, Derwent and Cow Green, to the far smaller direct-supply reservoirs at Whittle Dene. Some, like those three largest examples, are of comparatively recent construction, while many of the smaller flow-regulators, found at the very tops of river valleys, were designed and built under gas lighting and steam power. Initially, such developments were looked upon as a threat by the angling community, but it didn't take long for them to realise their potential. It quickly became apparent that these great lakes could, in fact, provide completely new angling opportunities in an age when traditional game fishing species, such as salmon and sea trout, were going into decline.

One of the first North East reservoirs constructed was Catcleugh, formed by damming off the River Rede about 10 miles upstream of Otterburn. Built in the 1890s, the construction process involved was highly labour-intensive and, at a time not long after the coming of the railways, two shanty towns grew up on either side of the remote dale to house construction workers, one nicknamed Newcastle and the other Gateshead. Indeed, when the Newcastle & Gateshead Water Company began building Catcleugh, one of the first projects they undertook was to lay a narrow-gauge railway connecting to the Wansbeck Valley or 'Wanney' railway line at West Woodburn in order that men, machinery and materials might be transported to the site more easily.

According to Michael W. Marshall, in *Tyne Waters: A River and its Salmon*, concerns were expressed at an early stage that eventually led to an order that a fish pass be incorporated into the new reservoir. Such plans, however, soon went awry: 'The directors of Newcastle and Gateshead Water Company offered to make a payment of £3,850 to the Fisheries Board of Trade to be excused from building the pass and the company promoted a new parliamentary bill for the building of the reservoir, this time omitting the construction of the fish pass. For technical reasons, the bill was not successful, but by then it had been decided to build a larger reservoir. A new act was passed, the fish pass abandoned and the Fishery Department got £3,850.'

Ultimately, the water company joined forces with the Duke of Northumberland to form an angling club that was still in existence at the time of publication of Marshall's book (1992). In the end, the only stipulation that Newcastle & Gateshead were held to was to install a grate to prevent the escape of the new angling club's stocked North American rainbow trout. Even this measure met with little success, however, as rainbows became a regular occurrence in the catches of trout and salmon anglers on the Rede.

More recently, according to the Wansbeck Angling Association, that river has suffered similar incursions from both Fontburn Reservoir, at the head of its tributary, the Font, and Sweethope Loughs, a former estate lake complex converted as a commercial trout fishery back in the early 1970s. Fontburn was built shortly after Catcleugh, much of the machinery and workforce transferring over to the new scheme after completion of the Redesdale project. It opened in 1901 for the Waterworks Department of Tynemouth, a company eventually incorporated into Northumbrian Water, but it would be many years before this reservoir realised its angling potential.

For although now a leader in the field, the old publicly owned Northumbrian Water Authority was quite inconsistent in its provision of stocked trout fisheries. While such venues as Fontburn and Kielder remained un-stocked 'wild trout' venues in the olden days, down in Teesdale, places like Grassholme had been run on a put and take basis for many years. Frank Johnson recorded that, as far back as 1975, there was 'an eight fish catch limit on the stocked reservoirs at Cow Green, Selset and Grassholme'. But for £2 a day, any angler that had bagged his total from one of these three was still allowed to move on and fish at the un-stocked Hury, Blackton or Balderhead, or even go for another eight-fish limit on the Authority's 2-mile stretch of the Wear at Eastgate.

While Fontburn was built to supply Tynemouth, as well as compensating for abstraction from the Wansbeck at Morpeth, Catcleugh contributed water to other parts of Tyneside, through a system of aqueducts connecting it to other former Newcastle and Gateshead reservoirs at Colt Crag, East and West Hallington and ultimately Whittle Dene. Each of these lakes, built only a few years after Catcleugh, has evolved into a trout-fishing venue in its own right, with Colt Crag and Hallington having, for some years, been controlled by the Little Swinburne, and Westwater Angling Associations, respectively.

Whittle Dene, meanwhile, was formerly a trout fishery of local renown, nostalgically remembered as the place he caught his first fish by local radio angling personality Railton Howes. Back in 1975, Frank Johnson wrote in his *North East Angling Guide*: 'The Newcastle and Gateshead Water Company allows fishing on eight reservoirs about 10 miles west of Newcastle. The season is from 1 April to 30 September and anglers pay 85p for a day permit, obtainable in advance, from the reservoir keeper's house. Boats are available at 15p an hour or £1 a day. Fly only, except after 31 May in the Great Southern Reservoir and 30 June in the Lower Reservoir.'

In the 1990s, after its owner merged with Northumbrian Water, Whittle Dene became equally renowned as a day ticket coarse fishery specialising in match-style angling for roach, perch and skimmer bream. Anglers can now have the choice of a form of reservoir angling not in vogue in the North East, back in the 1970s, while traditionalists are catered for on two lakes that are still preserved, privately, as fly-only trout fisheries.

The Newcastle & Gateshead Water Company remained independent of Northumbrian Water until well after the latter's privatisation, while its sister company, the Sunderland & Shields, supplied water to many other parts of Tyne and Wear from the newer 'flagship' Derwent Reservoir. Derwent was built in the mid-1960s and, at a vast 1,000 acres, immediately became the region's premier stillwater fly fishery.

Frank Johnson wrote that, 'Since it opened in 1967, Derwent has been stocked with more than 170,000 brown and rainbow trout and offers excellent sport to the fly fisherman. There is 7 miles of bank fishing available at £1.20 per day (season permits £34) and half price permits for boys and pensioners. Six fibreglass boats are available for hire, the charges being £2.50 a day for rowing boats and £5 for those fitted with an outboard engine.'

By the time the Newcastle & Gateshead and Sunderland & Shields water companies had become North East Water in the early 1990s, the fleet of boats was long gone. But when parent company Lyonnais des Eau merged with Northumbrian Water, after a second politically charged courtship in 1996, a rejuvenated fishery emerged. The fly-only rule was abandoned on the more accessible County Durham shore (it still remains in force on the Northumberland bank), replaced initially by a fly and worm sanction and, ultimately, by an all-encompassing any-method policy that was rolled out across almost all the most popular Northumbrian reservoirs. As part of

Anglers at Fontburn Reservoir.

Derwent's fortieth birthday celebrations, in 2007, Northumbrian Water introduced eleven 20-pound-plus rainbow trout as part of a blanket increase in stocking density that, combined with the relaxed bait restrictions would, it was hoped, vastly increase catch returns on the reservoir.

However, if we were to go back to basics and the *Piscator Non Solum Piscator* code of practice still embraced by many game fishermen, this is clearly an example of what some traditionalists regard as the problem with softening the rules. Many would argue that, while fishing for trout with a worm has its demands, lobbing a piece of legered multi-bait out and waiting for a fish to practically hook itself is definitely not playing the game.

Nevertheless, fishery managers would point to the increasing numbers fishing at reservoirs like Derwent as justification. That the changes followed hard on the heels of 2001's foot-and-mouth crisis, when most out of town fisheries were off-limits, was still greater reason – as something had to be done to cajole the public to come back. For my part, while I agree that multi-bait might just be missing the point, surely it's better for the sport of angling to see large numbers of people participating, especially families, and hence, the next generation of anglers.

In any case, it's not as though the 'fly only' rule has completely had its day. For those whose preference is, understandably, catching trout only by means they consider fair and sporting, there are still a few of Northumbria's reservoirs where this is the regulation. Those reservoirs where fishing is controlled privately, notably Colt Crag and Hallington, also retain the old code and it would be hard to find any of the many smaller commercial trout fisheries where bait fishing is permitted.

These much smaller stillwaters are now as common as the reservoirs themselves and can vary in can size from the 125-acre main Sweethope Main Lough, with its fleet of boats, to purpose-dug stew-ponds only a few yards across. And if the question of watercraft is an issue for the 'any method' reservoirs, then surely, on some of the smaller waters, the ease with which fish can be located is a very distinct advantage. Each unto their own.

In this competitive age of the 'leisure industry', adaptation is the key to survival. Like it or not, the future of commercial angling is the future of the sport as a whole and, with the tendency of certain reservoirs to even allow coarse fishing, nowadays there is something for everyone, bait and fly anglers alike.

CHAPTER SIX

Chasing Rainbows: A Brief Guide to Stillwater Trout Fishing in Northumbria

The discipline of stillwater trout angling can be subdivided into two distinct categories and, up here in the North East, we are lucky enough to be blessed with both forms in abundance. Reservoir angling, as we have seen, is done on very large lakes whose primary function is water supply, with many now also serving the secondary purpose of a fishery. The North East probably has more reservoirs than any other region in the UK and nearly all are managed as commercial put-and-take fisheries with rainbow trout, farmed and stocked, being the main quarry.

The other type of stillwater trout fishery – sometimes given the tag 'independent' – is in essence any lake or pond where trout fishing is allowed that is *not* a reservoir. These can comprise anything from the large 125-acre Sweethope Main Lough – an ornamental estate lake created in the eighteenth century – to more recent and much smaller fisheries such as Sharpley Springs whose purpose-dug lakes are only really oversized ponds. Some allow fishing from boats, some only from designated fishing platforms, while others allow anglers to wade in the margins; but apart from trout fishing, the only thing most have in common is that they are operated as small, independently run businesses. The independent fisheries thus provide a sort of variation less evident on the largely uniform armada of Northumbrian Water venues.

The explosion in the number of both forms of commercial stillwater trout fishery can be traced back to the privatisation of the Northumbrian Water Authority in 1989. Up until then, while stocked trout fishing had been provided on several of the old authority's Teesdale reservoirs, on others – most notably Fontburn in Northumberland – the opportunity to fish was only for what few wild trout were already in residence. Privatisation and the attendant commercialisation of all aspects of the water company's business changed all that. Within five years, Fontburn had become the North of England's premier trout fishery, with thousands of anglers a season fishing for stocked rainbow and brook trout, many of the former being introduced at double-figure weights.

Sweethope Lower Lough.

Similarly, while there were some independent stillwater trout fisheries before 1989, these were relatively few in number and it was only the growth in popularity of angling up on the reservoirs that led to the appearance of many more of these smaller, more intimate venues. But while Northumbrian Water was flexible to the point of eventually allowing multi-bait fishing on most of its fisheries, the independents remained true to the traditional values of the game fisherman, with fly-only restrictions engaging the minds of more conscientious and traditionally inclined trout anglers.

That's not to say that trout angling on the North East's reservoirs is necessarily easy, however, as in truth Northumbrian Water's fisheries are as difficult as the individual wishes to make them. From the relatively straightforward method of multi-bait fishing, whose simplicity defies description, through to the more demanding rigours of the good old-fashioned lobworm, there is provision for the angler for whom the complexities of fly fishing are all too much. The fly itself, meanwhile, is still as popular as ever, employing a similar yet subtly different knowledge of entomology (and other aspects of the trout's diet) to its equivalent on running water.

Northumbrian Water's most popular fishery provides scope for all these techniques and lying only 15 miles from Newcastle (turn off the A68 at Carterway Heads and take the B6278 towards Edmudbuyers) it comprises the perfect venue for the fly and the bait angler alike. With around 11 miles of bank split evenly between the devotees of each game fishing discipline, the 1,000-acre Derwent Reservoir has come a long way since its halcyon days as the region's first trout 'commercial', although some would contend that not all the changes have been for the best. The fleet of boats has long since gone and, although the fly-only rule was relaxed in the 1990s, it is at least still preserved on the northern bank.

This northern shoreline (the Northumberland bank) usually faces into the prevailing wind and as such is both a productive and a demanding place for fly fishing. Stillwater fly fishing is as challenging an art as its counterpart on the river and what fishing on running waters demands in the mastery of things like line drag, the requirement to cast to distance into a full-on gale can more than match! Hot spots include the Bay of Plenty, which is approximately 3 miles upstream of the dam, and Cronkley Bay (immediately above the dam), although most of that bank – which stands right over deep water – is worth a cast. In spring and late on, the patterns that have stood the test of time include the Black Fritz, Orange Fritz, Dawson's Olive, Cat's Whisker, Black Hopper, Diawl Bach, Viva and Olive Buzzer, usually presented on a sinking line. In the summer, the Northumbrian Water guide recommends the use of a floating line with a 'washing line' cast consisting of a Black Booby or Muddler on point, with a Diawl Bach or Bibio as dropper. On a bright day, it is best to revert to a medium sinking line matched to a weighted Dawson's Olive and a Zulu.

In stark contrast, the reservoir bed going out from much of the southern multi-bait shore (the County Durham bank) drops away only gradually, making it more suited to legering whenever the reservoir is full. Popped-up baits are usually favoured, such as the buoyant and highly scented 'powerbait' found on sale at the lodge below the dam. The out and out coarse angling technique of the block-end swimfeeder, filled to the brim with maggots (with a maggot on the hook, of course), will also account for fish by the barrow-load, whereas the slightly more traditional method of fishing a lobworm beneath a bubble float requires an extra long cast here, as at many points along this shoreline, even a hook-bait fished relatively close to the surface will snag on the bottom.

Fly fishing is allowed on this side of the reservoir as well, but on busier days the fly angler would be better off sticking to the north bank. When the opportunity does arise, however, there is at least one highly productive bay about quarter of a mile downstream of Pow Hill car park. Pow Hill and the south bank adjacent to the dam wall are the most productive areas for bait fishing. For Pow Hill, turn right along the B6278 towards Edmunbuyers from the lodge, and then right again at the fork in the village towards Blanchland (B6306). The turning for Pow Hill is clearly signed on the right about a mile along the B6306. Prices in 2011 are £23 for a day, £21 concessions, with an eight-fish bag limit, or £16 for the 'credit cruncher' ticket, which allows you to take four fish – telephone the lodge on 01207 255250 for more details.

Next on the list of Northumbrian Water's most popular fisheries is Fontburn Reservoir, located only a mile off the B6342, roughly halfway between where it crosses the A696 Newcastle to Jedburgh road (turn towards Rothbury) and Rothbury itself. Despite being one of the region's older reservoirs, Fontburn only actually opened its door to anglers in 1981, when the then Northumbrian Water Authority first allowed fishing for the large wild brown trout that were supposedly the progeny of ones that survived the damming process. In reality, Fontburn never contained any of those gargantuan brownies, but within a few years of Northumbrian Water's privatisation it was at least the home of double figure rainbow trout reared in the reservoir's own hatchery. Fontburn quickly acquired a reputation as the North East's premier big-trout fishery, a position it retained until Derwent came under the Northumbrian Water umbrella in 1996.

At 87 acres, Fontburn is one of Northumbria's smaller reservoirs, meaning that on busy days the sedentary bait angler might have a long walk to find a pitch that hasn't been taken, unless he or she arrives fairly early. Both multi-bait and fly fishing are allowed on both sides and, with the drop off being much greater than on Derwent's south shore, both legered and float-fished baits can be successful, depending on the time of year. There is a floating jetty for disabled anglers close to the car park on the northern bank.

The fly angler has the distinct advantage here that he can walk the entire bank relatively easily, taking a cast in convenient gaps between the bait anglers sitting in their 'pegs'. The areas on both sides towards the upper end of the reservoir (the very top end is a nature reserve) can then be good areas for a concentrated attack, as bait anglers tend not to like to walk too far with their heavier equipment, dropping into gaps or at the first available spot upstream of the last angler to have arrived. The most productive flies in the spring and autumn are lures (such as the Cat's Whisker or Dawson's Olive) or boobies on a fast sinking line, or wet flies on a slower sink. In 2009, as of early June, the best reported flies were Fritz, Black Fritz, Gold Headed Nymphs notably the Hare's Ear, Vivas, Black Hoppers and Daddies. In summer, the suggested combination is a Dawson's on the point with small wet flies on the droppers on a floating line with a slow retrieve.

The day tickets available at Fontburn, and the prices, are as for Derwent Reservoir. Telephone 01669 621368 for more information.

The longest established of Northumbrian Water's put-and-take trout fisheries, available to both fly and bait anglers, is Grassholme Reservoir, which is found high in the North Pennine Dales in what was originally the North Riding of Yorkshire. This area, which now comes under County Durham, was made famous in 1970s the by the ITV programme *Too Long A Winter*, about single female hill farmer Hannah Hauxwell, and it was here in 1984 that I took my own first reservoir trout – a brace of rainbows, in fact – after several unsuccessful attempts for Fontburn's mythical wildies!

Covering 140 acres, Grassholme is one of six Northumbrian Water reservoirs found in the Teesdale locality, with both it and the slightly larger

Selset located in Lunedale, and Balderhead, Blackton and Hury just a few miles south on the River Balder. To the north-west, on the Tees itself, is the largest of all the Teesdale reservoirs, the 770-acre Cow Green. Balderhead, Selset and Cow Green are all preserved as wild trout fisheries, each costing £10 for a day's fishing, while Hury is Northumbria's sole fly-only stocked trout fishery, costing £23 a day. Fishing at Grassholme comes under the same conditions and permit prices as those quoted for Derwent and Fontburn, and the reservoir is found by taking the B6277 from Barnard Castle (after crossing the River Tees, go straight ahead instead of turning left onto the A67). Carry on through Cotherstone, before turning left towards the reservoir at the village of Mickleton.

Grassholme responds well to both fly and bait fishing – and it can be worth hedging your bets. I well remember those two rainbow trout taken way back in the 1980s – on two 'last ditch' casts with a worm by the lodge, having thrashed the windswept southern margins all day with a plethora of wet fly patterns! The more fortunate (or competent!) fly angler will find that, in spring, the Cat's Whisker, Dawson's and Hot Tail Dawson's Olives, Bibio, Black Fritz and Olive Fritz all will catch fish on a slow sink line, with the compulsive gambler or heavy metal fan maybe opting for the Ace of Spades. As the water warms up, a floating line might be combined with an Orange Booby, and while either a Dry Mayfly or a Mayfly Nymph might also score, John Goddard's G&H Sedge comes highly recommended. The telephone number for Grassholme's Rangers and shop is 01833 641 121.

Last, but certainly not the least, of Northumbrian Water's premier reservoirs is the giant Kielder Water, the largest man-made lake in Western Europe, which opened for business in 1982. The trouble was, by the time it came on line, there was no business – the expanding chemical industry at the mouth of the Tees, for which Kielder's discharges had been intended, having gone into decline. Since its inauguration after seven years of arduous construction, Kielder has seen use of its primary service only once – during the extreme drought in the summer of 1995 – although it can produce up to six megawatts of hydroelectricity.

Fishing at Kielder is well established, with a trout fishery having been in existence on the upper part of the reservoir three years before the main section was flooded (Bakethin Reservoir, now a nature reserve and separated from the main lake during low water by a weir, was opened in 1979). Since the 1980s, trout fishing has taken place on the main part of Kielder Water and there are over 20 miles of bank for the angler to choose from, as well as a fleet of fifteen boats for fly fishing.

The north side of the reservoir is the less accessible shoreline, fringed all the way from Bakethin down to the dam by the equally vast Kielder Forest, which comes all the way down to the margins in most places. The most popular area on this bank is Hawkhope, which stretches for about a mile immediately upstream of the dam and is easily accessed from the car park at its northern end. The prevailing winds across and down the reservoir hit their furthest stretch of bank here and the only real issue with this excellent stretch, sitting as it does right over the deepest water in the lake, is that it's only really suitable for bait fishing, with high banks and trees that come right down to the water's edge. Still, there are few more exhilarating sights in angling than gazing across the waves towards the peak of Black Knowe with the while triangles of the sailing club's fleet 2 miles distant across the water at Leaplish, as your float starts to skate across the surface film with another rainbow, or maybe even one of the North Tyne's big browns, snaffling the worm!

Needless to say, on a lake the size of Kielder, fly fishing is best done from a boat in any case and these can be pre-booked by ringing 01434 251000. Early and late in the season, a medium sink line in conjunction with the Orange, Yellow and/or Green Fritz can be highly effective, as can the Nomad, Clan Chief, Greenwell's and Doobray. In summer, a switch to a floating line is most productive, with Jack Frost, Mini Vivas, Bibios and Black Snatchers being among the most favoured fly patterns. Kielder Water is approximately 12 miles west of the North Tynedale village of Bellingham and well signposted. Bellingham is situated on the B6320, which is reached by taking the A69 to Corbridge, then the A68 towards Jedburgh for 3 miles before turning left onto the B6318 'military road'. The B6320 begins on the roundabout immediately after the bridge crossing the North Tyne. The season for all Northumbrian reservoirs runs from 22 March to 30 September, although there are extensions in some cases.

Of course, the reservoirs listed up to now don't comprise all of those owned by Northumbrian Water; in fact there is even another – Scaling Dam, half way along the A171 Teesside to Whitby road – that is also part of Northumbria's own fisheries operation (telephone 01287 644032 for details). Otherwise, most of the other reservoirs in the region where fishing is allowed have at some time or other come under the control of angling clubs and associations and as such are open either to members only, or are operated like the independent trout fisheries, with a limited number of day tickets on offer. Notable examples of this are Catcleugh and Colt Crag, as described in the previous chapter, as well as Waskerley, near Consett, which offers up to five day tickets for worm and fly fishing (contact number is 0191 488 4873). Tunstall Reservoir in Weardale, part-leased by the Ferryhill & District Angling Club, is run along similar lines to Waskerley and information on membership and/or day tickets is available by calling 07825 951525.

Two of the North East's finest independently run reservoir trout fisheries are found at opposite ends of the region – Lockwood Beck on the North York Moors and Hallington Reservoir in central Northumberland. Both are run professionally and have many members and regular visitors. These fisheries offer the more discerning angler a combination of the kind of natural habitat more often associated with a reservoir environment, combined with the sort of angling experience not usually linked to the sometimes overcrowded multi-bait reservoirs elsewhere.

Lockwood Beck was formerly a part of the Northumbrian Water portfolio, but has for several years been leased by the Lockwood Beck Trout Fishery. In 2006, it was voted Britain's best trout fishery by readers of *Trout Fisherman* magazine and it is situated on the A171 between Guisborough and Whitby, approximately 10 miles closer to Teesside than Scaling Dam.

In common with a lot of the other 'independents', Lockwood Beck is fly (with barbless hooks) only, all anglers must come equipped with a landing net and priest, and wading is restricted in certain areas. The fishery is stocked weekly with brown and rainbow trout ranging from 1½ lb to double figures, and permits vary in price from £756 to £249 per season for adults (prices dependent on the number of fish the angler expects to keep, with unlimited catch and release), £60 per season for children aged between twelve and sixteen, £28 for day tickets, £20 for evening permits and £14 for four hour permits. A day ticket for juniors (aged twelve to sixteen) is £10 and under-twelves can fish for free with an adult (all 2011 prices). There are eight rowing boats available for hire on this 60-acre fishery and the season runs from 4 April through to 31 October.

The fly life at Lockwood Beck is similar to that on most of the Northumbrian Water-run reservoirs, with hatches of buzzers plentiful, and sedge, damsels, daddy longlegs and even swarms of flying ants providing food for the trout. It therefore follows that several of the patterns mentioned previously will catch fish here, with many of the other designs also worth a cast (the Black Buzzer, Hare's Ear and CDC Caddis Fly all being popular), and the use of sinking, medium sink and floating lines fits correspondingly to the varying water conditions over the course of the season. Bookings and further information can be obtained from the fishery manager on 07973 779527 or email info@lockwoodfishery.co.uk.

Hallington Reservoir is in fact two separate lakes separated by a causeway and, in contrast to nearly all the other reservoirs in the North East, the complex is part of a direct-supply operation, its water being pumped in from Catcleugh and Kielder and transferred by means of an aqueduct system to the nearby Whittle Dene Reservoirs. It is located a third of a mile from the village of Colwell on the B6342 (from Newcastle, follow the A69 to Corbridge, then the A68 towards Jedburgh. Turn right onto the B6342 about 10 miles up the A68 and take the first right turn after Colwell. From Hexham take the A6079 and cross straight over the A68 to continue onto the B6342). The fishing on both the east and west reservoirs is controlled by the Westwater Angling Club and is by fly only, with the season lasting from 1 April to 31 October.

Fishing at Hallington is run along similar lines to Waskerley and Tunstall reservoirs, in that the majority of the anglers that fish there are members of

Westwater Angling Club and day tickets are available only in limited number. In 2011, full membership cost £545, student membership £105 and junior membership £70, with boats available free of charge to members. Catch and keep limits are 6 trout per day, 30 in a month and 120 in a year, and additional catch and release is permitted in the same numbers. Members are allowed to bring a guest who can fish for £25 a day but no one individual is allowed to fish on more than two occasions a year. A small number of day tickets are available to non-members at £30 with an additional £10 for the use of a boat.

The most successful fly patterns at Hallington are again similar to those on most of Northumbrian Water's fisheries, with the very early season sport in 2009 having fallen to Pheasant Tail Nymphs, Black Buzzers and black midge patterns. Boat anglers caught mainly over deeper water on lures such as the Cat's Whisker. As spring progressed, Blae and Black, Butcher, Pupa and Bloodworm were all reportedly taking trout, with the Dawson's Olive and Appetiser joining the Cat's Whisker as the most successful lures. Eventually, as the summer weather took hold, the best of the fishing switched to smaller lures and traditional lake flies like the Invicta, Hare's Ear, Pond Olive Nymphs and Mayfly Nymphs. Further information about fishing at Hallington can be obtained by writing to Westwater Angling Ltd, The Clubhouse, Hallington Reservoir, Colwell, Hexham, Northumberland, NE46 4TT, telephoning 01434 681405 or emailing info@westwaterangling.co.uk. Westwater Angling's Website is: www.westwaterangling.co.uk.

From the privately run reservoir fisheries, the logical step is to move on to those larger private trout fishing lakes, none of which are in any way connected with the water supply industry. Some, like reservoirs, are nonetheless a part of the area's many river systems and one such example is Sweethope Loughs, located at the head of the River Wansbeck in Northumberland. As we know, Capabilty Brown created Sweethope in the eighteenth century as an ornamental estate lake and, to this day, the fishery is still controlled by local landowner, The Ray Estate. The contact for Sweethope is The Warden (Fishery Manager) on 01830 540349 and the loughs are found by turning left off the A696 Newcastle to Jedburgh road just north of the village of Kirkwhelpington and following this C-class road for approximately four miles.

Sweethope Loughs consist of two lakes, the larger of which, at 125 acres, is considerably bigger than near-neighbour Fontburn Reservoir and more than double the size of Lockwood Beck. As such, Sweethope Great Lough, with its fleet of twenty boats, is extremely popular among devotees of competition-style fly fishing and situated high on moorland, the winds that ordinarily sweep across this broad expanse of open water must give anglers adrift on the waves the feel of what it's like to fish on such exalted Scottish venues as Loch Leven.

Rainbow trout caught on fly at Sweethope Loughs.

The smaller Lough, at a mere 24 acres, is reserved for bank fishing, with wading permitted, and for many years the lakes have been stocked with hard-fighting rainbow trout of up to 20 lb. Both Sweethope Loughs operate over three four-hour fishing sessions a day, beginning at 8.45 a.m., 1 p.m. and 5 p.m. in the summer months, and each costing £17 (there is also a 'parent and child' ticket priced £20), with a three-fish catch limit. A £6 boat charge is extra. Eight-hour sessions are also available at £34, with six fish allowed to be taken on this ticket, but there is also a three-fish eight-hour session offered for £22 and an all-day sporting ticket, in which all fish caught have to be released, costing £17. A half-day season ticket for Sweethope, in which three fish a day can be taken, allows ten visits and costs £330.

Being the size of a reservoir, most of the fly patterns already discussed will work at Sweethope. Daddys and Fritz are particular favourites, with the Dawson's Olive accounting for many of the trout caught all year round. The bead-eyed Dawson's Olive, a variation with added weight, is very useful for locating fish that are running deep in colder weather.

About 20 miles south-west of Sweethope, still in Northumberland, Langley Dam is another example of a stillwater trout fishery with a long and rich history. Going back to my original assertion that we were now beyond the realms of the reservoir sector, this isn't strictly the case with Langley. Constructed at around the same time as Sweethope Loughs, this 20-acre lake was originally a river-regulating reservoir of sorts, built to supply water to the lead mines at Langley. It was this practice, of course, that wrought havoc on game fishing in the area throughout the nineteenth century, but nowadays Langley Dam only enhances the excellent range of options available in the South Tyne valley.

In actual fact, Langley wasn't even a game fishery to begin with. At around the time the South Tyne was still recovering from the effects of the pollution downstream, Langley was a coarse fishery operated by the Northumbrian Anglers' Federation. The lake was stocked with a variety of coarse species, including carp, some of which still survive and are frequently mistaken by anglers for large rainbow trout when seen cruising near the surface in the margins.

Similarly to Sweethope, Langley Dam offers four-hour (three fish), eight-hour (five fish) and parent-and-child (eight hours, six fish) tickets, but the hours you can start and finish are flexible. There are four boats, which must be pre-booked and the lake fishes well to both smaller traditional fly patterns and lures, with the Dawson's and Cat's Whisker being particularly effective. The fishery is located close to the A686 Alston road on the crossroads of the B6305 and B6285, which both meet the A-road nearby. The A686 itself meets the A69 only about five miles away at Haydon Bridge. Telephone 01434 688846 for more details.

Finally, we come to the smaller stillwater trout fisheries of the North East, most of which are fairly recent innovations that have nonetheless, in a relatively short space of time, built up a devoted following among the region's fly fishermen. The main attraction of these venues is that the large rainbow trout they also stock, unlike on larger waters, are never all that far away from the angler. Thus, customers at these stillwaters are given a very real sense of security that just isn't possible on the larger fisheries. It could be that their lucky day is much closer to hand than, for example, on the vast Derwent or Kielder reservoirs. The odds are stacked in their favour and the numerous photographs of ordinary anglers clutching double-figure rainbows displayed in tackle shops and on the sports pages of local newspapers bear testament. It also helps that the problems associated with casting a fly on large windswept bodies of water is rarely an issue at these smaller venues.

Nonetheless, you are seldom allowed to fish at these places by any method other than fly. Sharpley Springs (or Sharpley Waters), only open since 1997 and situated just off the B1404 between Houghton-le-Spring and Seaham, consists of four ponds, the largest of which is three acres, created by mining subsidence in the early 1990s. Unlike most of the larger trout fisheries (and in common with many other smaller ones), Sharpley remains open all year round as, in the 1990s, many of these venues challenged their legal obligation to comply with a close season that lasted half the year. They did this on the grounds that a dispensation for the much shorter close season for coarse fishing on stillwaters (not connected to rivers) was already in existence. By stocking rainbow trout (for which there was no official close season anyway) they were able to operate their own dispensation allowing them to trade – and fly anglers to fish – right through the winter. Information on fishing at Sharpley Waters is available by telephoning 0191 581 8045.

Another venue similar to Sharpley Springs is Aldin Grange Fishery, situated at Bearpark near Durham City. The two trout lakes at Aldin Grange (there are two other for coarse fishing as well) are strictly fly-only with barbless hooks, with the lower lake relatively shallow and the slightly larger upper lake a little deeper. Lure and buzzer fishing are the staples here, but fishing a dry fly can also produce the goods on a summer day when there's a light ripple on the surface. Small black patterns such as gnats and hawthorn flies, as well as daddies and hoppers, are most reliable, but look out for the hatches of caddis that often occur at around this time of the year. In such instances a brown sedge will frequently score. Aldin Grange is situated next to the Bearpark road out of Durham, immediately outside of Bearpark village itself (from the A167 between its intersections with the A690 and A691, turn down the hill towards Bearpark at the traffic lights and the fishery is on the left shortly after the road crosses over the River Browney). Telephone 0191 384 6090 for further details or visit www.aldingrangelakes.co.uk.

Other smaller stillwater trout fisheries are dotted throughout the North East and include Jubilee Lakes near Darlington (telephone 01388 772611), Beamish Fly Fishery near Stanley (07830 803336), Knitsley Mill near Consett (01207 581642), Witton Castle Lakes near Bishop Auckland (01388 488691), Sleekburn Water, just off the A189 Spine Road near Bedlington (01670 827030) and Hadrian Lodge Fishery near Haydon Bridge (01434 688688). They are all highly recommended – why not give one a try the next time you're thinking of a day's stillwater trout fishing?

CHAPTER SEVEN

Cast Against Type: The Unlikely Story of Coarse Fishing in the North East

If, like me, you're one of those anglers with a hankering for coarse fishing, you'll be pleased to hear that after six chapters dedicated to trout and salmon, it's now time for a bit of a chinwag about a branch of the sport all too often considered 'the poor relation'. Up here in the North East, in particular the counties of Northumberland and Durham, coarse angling is to this day still thought of in such disparaging terms by certain anglers. Old traditions die hard and the stock of the established trout and salmon angler on rivers such as the Tweed, Tyne and Wear still heavily outweighs that of those who target their prolific non-salmonid species, although it has to be said that attitudes have got a lot less antagonistic in recent times.

The presence or more to the point, the origins of non-game fish in these great rivers is still very much a thorny issue. Further south, the Tees and the Yorkshire rivers contain natural stocks of many cyprinid species, perch, grayling and pike; but, to the north, only the Tyne and Wear contain what are generally accepted to be indigenous coarse fish, namely gudgeon, roach, dace and chub.

The state of play on the Tweed catchment is less clear-cut, given that it lies 60 miles north of the Tyne, the river generally regarded as the natural limit for the northerly distribution of freshwater fish besides salmon and trout. Nonetheless, this river does contain several coarse species but, given its geographical position, not to mention the coarse-free status of at least two significant rivers further south, there is little doubt that, here, these fish are in fact 'aliens'. Moreover, clear lines of evidence can be found as to how these species came to be in such northerly flows in the first place.

The angling magazine *Waterlog* is frequently a mine of such information and, down the years, a couple of articles have proved very revealing of the origins of the Tweed's coarse fish. In 'Lords and Ladies' (issue 13), David Stirkazer described how, in 1860, a population of grayling introduced into an ornamental pond at Mounteviot escaped into the nearby River Teviot following a flood. These cousins of the native trout and salmon, equally adept at coping with swift flows, quickly adapted to the fast-running waters of their new home – 150 years on, the species now inhabits all but the very smallest rivers of the Tweed system.

In another piece ('Dainty Dace', issue 50), Fisheries Scientist Dr Derek Mills proposed that the dace, gudgeon and roach found in the lower reaches of the Tweed might be the result of less 'accidental' human activity at around the same time as the grayling's great escape. It would appear, according to Dr Mills, that salmon anglers of a bygone era used to use freshly killed dace, roach and gudgeon mounted on spinning tackle, in a kind of precursor to the devon minnow. These anglers would bring with them a stock of live specimens for their fishing and possibly released those left unused at the end of their stay. With the results of more deliberate introductions of non-indigenous species into other major river systems (e.g. barbel into the Severn), it's easy to see how these Scottish border dace, roach, gudgeon and grayling might have come to be so widespread.

And, insignificant though they may seem, these occurrences might also clarify the idiosyncrasies of several smaller Northumbrian streams. For example, the River Wansbeck, quite mysteriously, contains a population of gudgeon, while the Blyth (which borders the aristocratic Blagdon Estate) has a large head of grayling. In other places, the presence of such species is equally out of sync, but can be explained without any of the intrigue.

A little further south, on the River Derwent, records clearly confirm that grayling were intentionally stocked into this river in 1886 by our old friends the Derwent Angling Association – and the story goes that it was quite a technological undertaking! These fish were brought north from Yorkshire by train to Consett, whereupon they were transported to the river in milk churns. Apparently the fish only survived this last leg of the journey thanks to the quick thinking of a local lad, who took it upon himself to pour water from one churn to the next to keep the containers well oxygenated. Fiendishly clever!

However, not all such instances of local ingenuity have met with universal endorsement. Down on the Wear, far more recent examples of artificial propagation have only served to fuel an already heated debate.

A crucial facet in this river's vastly improved coarse angling has undoubtedly been the huge improvement in water quality following years of terrible pollution up until the early 1970s. Nevertheless, there was certainly more to the sudden re-emergence of numerous coarse species than just the cleaning up of the river. Several old hands on the Durham scene can pinpoint the exact date the Wear's now thriving dace population suddenly exploded, with fish brought in clandestinely from the Tweed, from where they were apparently to be 'thrown up the bank' in one of the trout and salmon brigade's infamous clearances. Likewise, none other than an Environment Agency bailiff once told me the name of the cul-de-sac in Chester-le-Street next to which the river's first barbel are believed to have been introduced. These would appear to have come, rather by an act of man than of God, from one of the North Yorkshire

Summer coarse fishing on the River Wear at Durham.

rivers to which they're indigenous, and similar introductions at around the
same time may have rejuvenated the river's natural chub population.

Yet, if you were to go back thirty years, the official angling literature of the
day would have had you believe that the Wear was only inhabited by native
brown trout and the returning stocks of sea trout and salmon. Such were the
prejudices back then, in an area where the tweed-suited gentry still regarded
coarse angling as a peasant's pastime and they appeared to be using the
disappearance of coarse species during the Wear's many decades of pollution
as proof that they were never native to the river in the first place. Their
outdated and scientifically inaccurate assertions that coarse fish had only been
introduced illegally was entrenched by those very real incidences of unlawful
stocking, even though this was only, for the most part, restoring that which
had simply been destroyed by human activity.

The debate still rages on, although intellectually challenged suggestions
from certain quarters (such as having the entire river system poisoned to
remove all the coarse species completely) are now, thankfully, a thing of the
past. In recent years, chub, dace and grayling have all been officially stocked
into the Wear by the EA on the pretext that they are indigenous and, indeed, a
British record dace was caught from the river as recently as 2002.

The stillwater coarse angling scene in Northumberland and Durham is unlikely
to be the setting for any records, but the enthusiasm with which its devotees now
practise their art is a match for any hallowed pond, pit or lake further south. Like
such fishing on rivers, stillwater angling for species other than trout was also slow

to gain recognition in the North East, even though Robert Blakey wrote as early as 1854, 'The lakes in the north of England are full of perch and private ponds in which jack pike are preserved generally abound with them.' Is it reasonable to assume, then, that in the decades before rainbow trout were introduced from America, aristocratic practitioners of the sport in Northumberland used to fish for perch and pike on such estate lakes as Sweethope?

Even if this were the case, modern stillwater coarse fishing in the North East didn't begin to take off until well into the twentieth century and it wasn't until 1971 that the Big Waters Angling Club became the first to offer public access. Born of an initiative by the Northumberland Wildlife Trust and Tyne and Wear County Council, this new club sought to provide quality coarse angling opportunities to locals, centred, to begin with, on the aptly named 22-acre North Tyneside lake.

Cast against type, Big Waters' grand opening had all the pomp of the first day of the salmon season, with Lord Ridley of nearby Blagdon Estate, no less, making the inaugural cast. But despite such a dignified scene (or possibly because of it), neither His Lordship nor any of the other anglers could make contact on that first day – something that was to become a frequent predicament down the years on this large and very challenging water.

At least by 1976, one angler was enjoying great success on Big Waters' vast acreage. In his article 'Phoenix from the Flames' (*Waterlog*, issue 45) the teenage Geoff Clarkson recalled great times that scorching hot summer, 'filling my keepnet with perch on many occasions'. On the face of it, these catches may just seem another youngster's vast haul of stunted stripeys, but Geoff's fish averaged half a pound and frequently ran the whole sixteen ounces – a specimen by 1970s Northumberland standards and vindication for Blakey's observations of one hundred and twenty years before.

'Easier' stillwater coarse fisheries would eventually come to pass in our northern climes but, for the time being, equally demanding venues were the norm. Further up into Northumberland, Bolam Lake provided inexpensive fishing and, being more featured, easier pickings to local junior perch anglers, but the pike were a much harder nut to crack. Back on Tyneside, the Leazes Park Angling Association was formed in 1976 out of the need to oversee carp fishing on a three-acre lake in a Newcastle park. For over a century, this place had been a Mecca for people seeking an escape from the bustle of the nearby city centre, but even a constant scrimmage in the summer months couldn't prevent an impressive list of members joining the 'Twenties Club' over the coming years!

At around the same time, down the road at Durham, a local angling club acquired the rights to a former brick pond close to Low Newton, high above the River Wear on the city's northern outskirts. It offered members of Durham City Angling Club semi-rural peace and tranquillity in the style of Big Waters, combined with a surface area only slightly greater than that of Leazes Park

Lake. Like the other two, a mixed stock of coarse species was present, but also like the others, the pond's carp would soon become the main attraction.

Two decades later, as carp angling became the poor man's salmon fishing the nation over, 'commercial stillwater coarse fisheries' started springing up everywhere. These were day ticket fisheries where membership of a club was not required and served the dedicated and casual angler alike. Everywhere, that is, except Northumbria, where carp still lagged some way behind commercial stillwater trout fishing, in whose ponds, lakes and reservoirs stock-reared rainbows could reach (or more often were introduced at) weights well in excess of that of big North East carp.

Still, the coarse commercials would eventually come and Angel Fisheries (a purpose-dug set of ponds near Birtley) and our old friend Whittle Dene have blazed a trail in this traditional heartland of trout and salmon. Further south, Woodlands Lakes in North Yorkshire has been in existence nearly twenty years and attracts custom from all over the North of England. In many parts of the country, this branch of the sport is now seen as the future of angling and those exclusive game fishing interests had better watch out – accessibility is the key!

Down in Yorkshire, river angling has always been mainly about coarse fish, although looking back through historical literature, it can be almost as difficult to nail down hard facts about fish stocks in this county's streams as it is for the ones in Northumbria.

The problem is that back in the nineteenth century, books were written for a southern middle-class clientele, the quintessential Middle England that late twentieth-century politicians sought to woo. Back then, the average Middle Englander could enjoy fishing for anything from gudgeon to barbel on rivers close to home – the Thames, Kennet and Trent, and, from the late 1800s, even chalk streams like the Hampshire Avon. What they couldn't always rely on was the trout and salmon fishing, but with unprecedented travel opportunities provided by the railways, the option to head north became a common leisure pursuit for those anglers with the wherewithal to do so. Thus guidebooks about or including the north of England either brushed over or completely ignored any coarse fishing that was on offer, the authors seeking to stereotype all northern angling in a guise that suited their southern readership.

In 1854, Blakey wrote, in his tome *Angling; Or, How to Angle and Where to Go*:

Angling in England and Wales is to be viewed in a somewhat different light from Angling in Scotland and Ireland. In reference to England in particular, anglers may fairly enough be divided into two classes; the one pursuing the bottom fishing, and the other, making the salmon, trout and pike their chief source of amusement; the one class principally confined to the metropolis and its extensive suburbs, and the other located on the banks of numerous rivers and estuaries at the more distant parts of the kingdom.

In his chapter on fishing in England and Wales, the author thus described in some detail the quality of barbel, perch, chub, roach, dace and ruffe fishing at such Thames locations as Battersea Bridge, Putney and Brentford. Later on, as the narrative moved up as far as the Swale, equally prolific stocks of precisely those species were ignored, Blakey simply commenting, 'The same kinds of flies that answer for the Wharfe will do here.'

By the end of the Victorian era, trout and salmon fishing had become so elite that even angling guidebooks written by local authors were falling into the trap. *The Yorkshire Angling Guide – A guide to the whole of the fishing on the Yorkshire Rivers*, penned in 1894 by Leeds angling journalist Tom Bradley, was a comprehensive manual encompassing every river in the county open to angling. Yet even compared to its relatively threadbare modern equivalent, *River Angling in Yorkshire*, published almost a hundred years later, detail was thin with regard to the county's extensive array of coarse fishing.

Bradley preferred to concentrate his efforts on such information as accommodation, nearby inns and railway connections – those more inaccessible locations being in the dales, where the pick of the trout fishing was to be found. Of the six and a half pages in the book dedicated to the Swale, only two concerned themselves with the 40-mile section of river below Richmond, 'the deeps' as he referred to it, that part where coarse species begin to predominate over trout. Interestingly, while the author did actually detail those coarse species present, 'pike, chub, roach, dace and eels', he chose to omit the one for which the river became most famous – the barbel.

In describing the Ure, Tom Bradley was even more circumspect, confining information on this fine mixed river to the quality of its grayling fishing at Hawes and pike fishing at Boroughbridge, and like for its sister river, his detail on the Ure was heavily weighted in favour of the upper reaches. But yet again modern literature paints a more cosmopolitan picture. Harking back to Derek Mill's *Waterlog* article 'Dainty Dace', Dr Mills recalled catching chub, roach, dace, perch, trout and grayling, as well as pike, during his childhood fishing forays at Boroughbridge in the 1940s. He also wrote of barbel and bream, thus completing the Ure's complement of indigenous fish.

For, unlike on the Wear and Tyne, cyprinid species are not on the boundary of their natural range in the Yorkshire rivers. The Vale of York, into which the Swale and the Ure flow, is a warmer place than the coastal plains of Northumberland and Durham, shielded from the harsh winter winds off the North Sea by the Cleveland Hills and North York Moors. The middle reaches of both rivers, as well as their languid offspring, the Ouse, are deeper, slower flowing and more extensive than any of their North Eastern counterparts, making them the perfect habitat for coarse species to thrive. The vale lies over the terminal moraine left by the southernmost extent of the ice sheet during the last ice age, meaning the riverbeds consist of alluvial silts ideal for species

A laid-back barbel angler in action on the River Swale.

like barbel, more sensitive to the winter chill, to sink into and from which dense weed, essential to spawning, can grow.

The simple fact is that on the Swale, Ure, and all the other tributaries of the River Ouse further south, coarse angling is predominant. Trout fishing does still prevail in the dales, where coarse species cannot exist, but elsewhere the tradition has always been to catch their less noble brethren, whatever the Victorian guidebooks might have suggested.

Last but by no means least, the River Tees falls somewhere in-between the bipolar natures of the true Yorkshire rivers and its more northerly Northumbrian counterparts, as might fit the traditional boundary between Durham and Yorkshire. Indeed it has even shifted allegiances on several occasions! While the administrative borders of County Durham were moved south-west in the early 1970s to annex much of southern Teesdale and the entire upper river, the Tees itself defected from the old Northumbrian Water Authority to the Yorkshire Dales division of the Environment Agency upon that authority's acquisition of the area's fisheries in the 1990s.

Coarse fish have always existed in the middle and lower reaches of the Tees and its tributaries. Bradley, while pandering characteristically to the upper river's charms, observed just before the turn of the twentieth century that there were grayling, chub, dace and pike in the main river at Croft and pike, chub, roach and eels in the Skerne. Further downstream, just above the tidal limit, he reported, 'Trout and grayling all over. Dace are numerous, also chub, which run large, and a few roach and perch.' It may have been this propensity to produce large chub that resulted in the river gaining a British record of 8 lb 10 oz in 1994, one hundred years after Bradley's observations.

And a degree of coexistence, less evident on other Northumbrian rivers was also at play, even in the late nineteenth century. Bradley recorded that below Dinsdale, 'All legitimate baits' were permitted and that a close season between 1 October and 31 March was in place, 'except for coarse fish'.

Half a century later, nothing much had changed, with Brian Clarke recalling a similar situation in a *Waterlog* article called 'Snaffle-on Tees':

> The near water (at Croft), tangled with branches and roots was the haunt of chub. Further out there were dace and trout. In the middle, too far to cast, there were, in those faraway days when imagination ruled, who-knows-not-what. Monsters for sure.
>
> I fished in two ways then: float or leger. On those sleeking weak tea Northern waters that was all most lads did. All most adults did. The idea of free lining or fly fishing or the like would not have occurred. Anyway fly fishing was for toffs.

With the late twentieth century came inevitable changes to such idyllic scenes, with results good and bad depending on your point of view. Brian Clarke wrote of the Tees upstream of Croft as 'a lovely light-laundered piece of water that rumpled its way downstream before the Ministry of Truth stepped in with "improvements"'.

And later improvements on a far larger scale were seen in the early 1990s following the construction of the Tees barrage across the estuary at Stockton. This project had the dual effect of preventing natural tidal backwash into much of the lower Tees and impounding several miles of former upper estuary as a slow-running freshwater canal. The principal purpose was the provision of amenities below the obstruction, but the restriction of brackish estuary water into the river above it inadvertently improved fishing for those coarse species already in the lower reaches, although other environmental factors have been less positive.

The lower Tees now provides a type of coarse fishing never previously seen north of York, with large bags of indigenous species showing, as well as that great lover of the more sedate flow, the bream. In the still free-flowing middle roaches around Croft and Dinsdale, the chub, dace and grayling have been joined by the barbel – a possible result of introductions by the same hand that gave the Wear its stock.

However, the changes along the course of the River Tees are viewed, this fulcrum of the North East rivers is now the definitive example of a Pennine spate river: a game fishery in its upper reaches, a mixed fishery in the middle and a coarse fishery at the bottom. Surely the way all major Northern rivers ought to be!

CHAPTER EIGHT

Trot Down, Bag Up: A Brief Guide to Coarse Fishing in Northumberland, Durham and North Yorkshire

It would be wrong to suggest, despite the huge increases in the availability and accessibility of coarse angling on river, pond and lake in Northumberland and Durham, that these two counties can offer anywhere near the quality and range provided by similar fisheries down in North Yorkshire. Geography and climate alone dictate that whether your preference is for carp, tench or bream from a stillwater, or chub or barbel on a river, the likelihood is that you're going to do better the further south you go. It's a similar story with silver fish (and perch) in both still and running water. And while specimen pike are present in several stillwater fisheries in these two counties, would such leviathans be found in any of their rivers? With the exception of the Tees and the Wear, the answer is a resounding no – quite simply because, apart from these two, the availability of fodder fish (mainly roach, dace and perch, as well as small trout) is insufficient anywhere else for pike to pile on the pounds in the way they can in a contained environment.

Nevertheless, what is about to follow wouldn't make for a particularly interesting chapter – not least for any reader from the far north of the region – if all the fisheries mentioned were in Yorkshire! Therefore, this is a guide to several of my favourite coarse fishing venues – stillwater and river – starting in Northumberland and heading south to the heartland of North East coarse angling down in North Yorkshire.

We begin in pastoral Northumberland with a venue whose influence on my own embryonic angling career makes it the coarse fishing equivalent of the river Wansbeck at Morpeth. The 25-acre Bolam Lake is situated just 10 miles west of the market town, and the former estate lake has all the character (if not quite the size) of Sweethope Loughs. Gifted to the County Council in the early 1970s, it became a country park whose popularity has increased greatly over the years, and it's now a firm favourite with visitors from all over the region. My arrival on Bolam Lake's angling scene came just a few years after it opened and, from the late 1970s to mid-1980s, its stocks of perch and pike provided a welcome alternative to the Wansbeck's brown trout.

A North East perch of a little over 2 lb.

Thankfully, Bolam Lake has changed little in the last thirty years, except for the vast numbers of tourists now visiting the site in the summer months. Like the trout in the Wansbeck, the fish I caught here back in the 1970s and 1980s were never all that big – mostly perch that occasionally touched a pound in weight and the occasional jack pike. Today the size of the fish is pretty much the same, making this more the sort of venue for enjoying your fishing, rather than a place to break any records.

And unlike many an un-stocked water, the numbers of fish haven't changed much here either, the perch having even survived a breach in the dam in the early 1990s, when the resultant drop in depth reduced their favoured habitat, the margins, to mere mudflats. Mercifully, the perch are still prolific and, besides the jacks, there now appear to be more large pike in the lake than ever. There are also, apparently – if you fish near the reed beds towards the top end of the lake (the uppermost area is a wildlife sanctuary) – even a few roach and golden orfe in residence, but in all honesty whenever I've fished a maggot or worm in these places (cereals aren't allowed), I've always ended up catching perch!

There is a very real chance of bagging-up with the perch at Bolam Lake in the summer months, and for this reason it can be the ideal place to take young anglers and keep them interested. If you can locate a shoal and regularly catapult pouches of loose-fed maggots, twenty or thirty fish at a half-hour sitting, or less, is entirely possible. Just remember to scale down your tackle and hook sizes accordingly and watch for telltale boils and swirls near your taking zone – or you might find the next small perch suddenly 'gets big' on you as Mr Pike intercepts!

It also pays to move pitches regularly at Bolam Lake, as the shoals of perch seem constantly to be on the lookout for new places to feed (and for pike) and the action will never remain in the same place for long. Hot spots include the corner of the lake immediately downhill of the main car park and visitor centre (facing the small island); the jetty on the far side near the boathouse wood car park; the entire north-west-facing shore of the big island (looking out onto the lake) and the island cum peninsula on the opposite side – all of which are accessible from the main footpath going round the lake. Check the depth of the water you're fishing over (at the latter 'hotspot' it will likely take a good cast to locate the fish) and float fish maggots or worm.

The pike are an altogether different proposition, of course, although I have caught smaller specimens that took a maggot or worm intended for the perch – and even one or two that took a perch! Real pike fishing at Bolam Lake has to be done using a dead-bait (live-bait and spinning are prohibited), with presentation entirely at the angler's discretion. Static, popped-up, float-fished or wobbled baits will all work equally well on their given day, but an attack may need to be 'provoked' in the colder months. As dead-bait can't be caught from the lake itself, sea fish such as smelt or spratt are generally the bait of choice (available frozen from most tackle shops), with mackerel and eel-sections also taking fish.

A few important things to bear in mind if you're relatively new to pike fishing are that you'll need to use a wire trace, a proper rod (1½-lb test curve or stronger), as well as strong line – and the ability to remove treble hooks from a fish (take a pair of long forceps, a glove and something with which to keep the fish's jaws apart humanely) is not only a matter of principle for most pike anglers, it's also the rules at Bolam Lake. In fact, regular pike anglers don't recommend that you try fishing for them without the help of someone reasonably experienced the first few times – just in case there do happen to be any problems.

Bolam Lake's pike aren't regularly taken in weights much into double figures, but the odd twenty (20 lb) has been reported in the past decade or so. The best places to fish are off the big island, particularly the end peg where a channel runs between the two islands at the dam end. Pike patrol through here and along the deeper channel between the islands and the dam, so the dam wall itself can be worth a try. It also goes without saying that in most of the places where you catch small perch, you will also find pike, although these won't necessarily be the bigger ones – and anywhere near a reed bed is always worth a go.

Bolam Lake is open during the traditional coarse angling season of 16 June to the following 14 March inclusive. Day/season tickets are available from Belsay village shop, telephone 01661 881207, priced £6/£24 (2005 prices).

1. Fishing a summer spate on the River Ure near Ripon.

2. Autumn fly fishing water on the River Derwent.

3. The River Wansbeck at Scots Gill.

4. Classic early season fly fishing water on the River Wear at Croxdale.

5. A winter chub from the River Wear.

6. A classic barbel swim on the River Swale.

7. The author with a County Durham common carp of 14 lb.

8. A 7¼-lb River Wear barbel.

9. A tench of around 3 lb caught at a North East stillwater.

10. A trio of Kielder Rainbow Trout caught on worm.

11. Sunset at Grassholme Reservoir.

12. Winter grayling water on the River Wear near Durham.

13. Cambois Beach, looking across the bay towards Newbiggin.

14. Lockhaugh Meadows in the valley of the River Derwent.

Left: **15.** The River Wear in midwinter. Amid the tangle of branches and roots lurk big chub!

Below: **16.** River coarse fishing in high summer.

17. The River Swale in summer. This swim, above the bridge at Great Langton, fishes well for chub and perch, and in spate, for barbel.

18. The River Tyne downstream of Hexham.

19. The River Wear at Croxdale in high summer.

20. The haphazard angler's stillwater pitch.

Right: **21.** The River Swale
looking downstream towards
Richmond Castle.

Below: **22.** The Swale at
Easby has many inhabitants.

Left: **23.** It's not only fish that like casters!

Below: **24.** Once a popular haven for pleasure craft, the shallow estuary of the River Wansbeck is still a productive mark for flatties and the occasional bigger fish.

25. Fontburn Reservoir, looking towards the valve tower and dam wall.

26. Weir pools on any of the Yorkshire rivers, such as this one on the Swale, are a likely hotspot for barbel.

27. Local knowledge can be a vital part of river fishing. At first glance, this stretch of the Ure looks to have potential, but the real hotspots are the faster water immediately up and downstream.

28. Ideal summer conditions for the river coarse angler: the river is coloured with the level dropping-off following a spate, and the weather overcast and muggy.

29. Returning a tench safely to a County Durham pond following its capture.

30. The River Wear at Croxdale – an excellent winter chub fishing venue.

31. A former weir on the River Wansbeck near Mitford. The structure was destroyed by the same floods that devastated the nearby town of Morpeth in September 2008.

32. A typical example of a small commercial trout fishery.

33. Ready to go! All the essentials (minus waders) required for a successful day's river fly fishing.

Alternatively, ring 01661 881234 or email bolamlake@northumberland.gov.uk for more details. The lake is found by taking the A696 Newcastle to Jedburgh road as far as Belsay and turning right (the correct turning is 'second right' by turning across the junction of the B6524 to Morpeth). Follow this minor road for about three miles until you reach the lake. Alternatively, take the B6524 which leaves the A197 at Morpeth (next to the golf club) and follow this road for five miles to the village of Whalton. At Whalton, turn right beside the Beresford Arms and follow this minor road for a further five miles until you reach the lake at a T-junction.

Next on the list is Whittle Dene, a direct-supply reservoir (as described in the chapter 'Reservoir Dogs') owned by Northumbrian Water, and situated about ten miles west of Newcastle. It consists of a set of small lakes and is situated on the junction of the B6318 'military road' and the B6309 at Harlow Hill in Northumberland.

For almost a decade, these formerly stocked trout fishing lakes have been allowed to go wild and their natural stocks of roach, perch, dace and gudgeon have now taken over. This is very much a venue for the angler who likes to bag-up in the match fishing style (most of the fish are relatively small), and double figure nets are a weekly occurrence. In 2006, skimmer bream were also introduced into the Lower Lake, with these stocks now helping to make Whittle Dene an excellent mixed coarse fishery.

There are four lakes to choose from: the smaller Northern and Western Lakes are uppermost on the site; the largest, the Great Southern Lake is lowest; and the Lower Lake in-between in terms of both size and location. All match-style coarse fishing techniques can be successful here. The float (a light waggler) will score in colder weather with a maggot fished hard on the bottom over a bed of ground-bait, while in summer fish will often be located higher up in the water.

The swimfeeder, meanwhile (use an open-ended one with ground-bait and a 'block-end' for particles such as maggots or hemp seed) will often pick up the better stamp of fish from the shoals of roach and bream. Just to be sure, the Northumbrian Water guide recommends the use of two rods, one cast long and one short, and remember that the use of a marker – a sliding stop knot tied onto the line (see Chapter Twelve), an elastic band around the spool or the reel-line itself fixed into the line-clip on the side of the spool (at the correct casting distance) – is essential, when combined with a far bank reference, to position the hook-bait in the right place every time. Another tip, when using particles in a ground-bait mix, is to allow the rig to settle for a few seconds before pulling it back towards you a foot or two to place the hook-bait in the middle of the ground-bait mix.

The most effective technique of all in this type of coarse angling is the pole set-up, employing a fixed line connected by shock-absorbing elastic to an

extendable (and retractable) carbon-fibre fishing pole. The idea is to place a
bait in exactly the same place every time in a way that isn't possible by casting.
In skilled hands this is a deadly technique for fishing tight to a feature such
as a shelf – common on reservoirs such as Whittle Dene. Hemp, bloodworm
and caster are the hook-baits of choice for this method in spring and autumn,
with maggot and caster being best in summer. The technique relies on accurate
loose-feeding – little and often.

The season at Whittle Dene runs from 1 March to 30 November in each
year – to get there, leave the A69 westbound at the sign for Heddon-on-the-
Wall (B6528) and, once in the village, turn right onto the B6318. Follow this
road for 7 miles until you reach the reservoir complex.

Just a few miles north of Whittle Dene, Dissington Pond near Ponteland is
another small stillwater, this one run by that well-established coarse angling
club, Big Waters A.C. This is BWAC's premier match fishing venue and, as
such, it responds well to all the techniques outlined for Whittle Dene – but
there's much more to this place than just its silver fish.

Being a former limestone quarry, Dissington is a very deep venue, varying
from 8 feet at the shallow end to over 16 feet at its deepest. Thus, when fishing
for the pond's carp and tench – generally too big and powerful to be fished for
on a pole – here you can make the most of that old-fashioned yet devastatingly
effective technique, the slider-float method, one of the most exciting ways of
coarse fishing.

Many people still prefer the use of a float on small stillwater venues, for
reasons of aesthetic appeal, as well as the fact that you can look round without
having to keep your eyes glued on a quiver-tip. The problem with conventional
float fishing on a very deep stillwater fishery, however, is that it makes casting
extremely difficult if you are to place a bait right on the bottom. Trying to cast
16 feet of slack nylon with a hook on the end, even to a position right under the
rod tip, is not the easiest thing to do from a peg surrounded by long grass and
other potential bankside snags. The answer is the slider float (described in much
greater detail in Chapter Twelve), which permits you to cast with only a few feet of
spare line, but will allow a heavily shotted rig to slide down through the eye of the
float and cock at the correct depth – providing you've plumbed it up correctly!

A heavily shotted float such as a bodied waggler is usually required to allow
the fairly heavy line you'll need (5 lb+) to slide freely, but as tench usually give
fairly solid bites when they're in the mood, this shouldn't be an issue. In this
instance, a conventional hook-bait such as sweetcorn or luncheon meat would
normally apply, but the heavily fished nature of this small pond means that a
degree of ingenuity may be needed. Again, this method depends on the 'little
and often' loose-feeding principle, although, with tench and carp, it doesn't
hurt to introduce a bed of either particles (hemp seed is particularly effective)
or ground-bait before you start fishing.

Dissington Pond has no close season and is found by taking the A696 Jedburgh road from Newcastle, passing the airport and continuing right on through Ponteland village and out the other side for about half a mile. On the left there is a turning signposted for Dissington and Dalton – a road called Limestone Lane – this is the clue! Turn into this side road and almost immediately, on the right, there is a turning into a farm track with a gate on its left. This is the entrance to the Dissington Pond car park. Big Waters Angling Club doesn't sell day tickets for this venue, but details on how to become a member can be found at www.bigwatersanglingclub.dreamstation.com.

Moving south into County Durham, we come to another top venue where a mixture of carp, tench and silver fish provide the mainstay of the fishing. Brasside Pond is that same former brick pond near Durham described in the previous chapter, but over the past five or six years it has undergone something of major transformation.

Brasside came under the ownership of Durham City Angling Club in the 1970s and by the mid-1990s, the club had drawn up plans for an expanse of water that, at 10 acres, was really more of small lake. The pond, in its original state, had both a deeper section, almost the depth of Dissington in places, and extensive shallows that made up most of the opposite end. DCAC's scheme involved building a causeway to separate the two main parts of the lake, as well as the construction of two completely new waters – the first an extension of an existing backwater and the other a completely new canal-style water for match fishing, dug out of adjacent marshland. The new complex now comprises one of the best club-owned coarse angling venues in the North East and caters for match, pleasure and specimen anglers alike.

For my part, I've always favoured the tench fishing over the deeper water at the bottom end of the lake, using a similar technique to that at Dissington. This part of the old pond now comprises the 'specimen water', as, after the new complex was completed, most of the bigger carp and tench were netted from the shallower 'pleasure lake' and permanently rehoused over in the deeper water. The bank to the left of and perpendicular to the pond's car park comprises a 100-yard-long trench which, with depths of up to 10 feet in places, is ideal for this sort of fishing from each of its nine pegs.

There is also considerable attention given to carp fishing, of course, and with the fishery record having passed into the twenties in the last decade or so, there are many who regard Brasside as the premier club-run carp fishing venue in the area. Unsurprisingly, the usual technique employed is a variation on the bolt-rig/method/boilie approach, employing optonics and a liberal amount of 'spodding' (I'm not knocking it – this is how I caught my own biggest Brasside carp!), and there's even the occasional bait-boat seen cutting through the waves.

Nonetheless, several more adventurous souls are occasionally seen casting a floating crust out into the middle of a fleet of free-offerings, hoping to catch a mid-summer specimen while the leger-lads struggle for a run. This technique has to be one of the most exhilarating ways of catching carp and it is described in detail in Chapter Twelve.

Brasside Pond is located opposite the main car park of Frankland Prison (don't be put off – no one ever escapes). From the roundabout on the A167 at Pity Me (just north of Durham, 3 miles south of Chester-le-Street), follow the road signposted for the Arnison Centre. Go straight ahead at the first roundabout then turn right at the next. Turn left at the next roundabout and go down the hill for about a mile, passing under a railway bridge and on through Brasside village. After the road to Finchale Priory turns off to the left, continue on down this narrowing lane for 100 yards and the gate for Brasside Pond is the first turning on your left. Durham City Angling Club does not sell day tickets but information about membership is available on their website at www.durhamanglers.co.uk.

In addition to Brasside Pond, DCAC is one of the principal angling clubs to hold bank-space on the middle section of the River Wear, with access to four separate stretches between Croxdale and Chester Moor, two of which we have already talked about in the trout fishing section. All offer excellent opportunities for the coarse angler, with barbel, perch, pike and even carp present in certain locations, and chub, dace, roach, gudgeon and grayling commonplace throughout.

With regard to coarse fishing, the Wear is a textbook example of a classic British river, rising in mountains, with all five of the main ecological zones recognised by the all-round angler – trout, grayling, barbel (chub), bream and tidal. Like all North Pennine catchments, the Wear is a spate river, beginning life less than a mile from the source of the South Tyne on the Durham–Cumbria border. But unlike either the Tyne or its other near neighbour, the Tees, the lower reaches of the Wear take inflow from hills composed of magnesium limestone, meaning its water ordinarily flows gin clear, in contrast to the peaty weak tea colour that the others usually carry.

From the beginning of the Wear 'proper' at Wearhead, the river flows some 25 miles eastward, passing St John's Chapel, Stanhope and Wolsingham en route, with little or nothing in the way of serious coarse fishing to speak of. By the time it reaches Bishop Auckland, however – about ten miles upstream of Durham – the river has lost most of its upland wildness, the surrounding hills slip down into pastoral farmland and grayling, chub and barbel live alongside the better nourished brethren of Weardale's diminutive brown trout.

The river changes direction here, forced into a north-easterly direction by that same magnesium limestone ridge that fines its lower waters, then north from the southern outskirts of Durham City up to Chester-le-Street. By now, the Wear is very much into its 'bream zone', although just about every species imaginable is actually present.

It is on this section that we find the lowest of DCAC's club waters, Chester Moor, just upstream of Chester-le-Street and accessed by turning left onto a farm track from the southbound carriageway of the A167 Chester-le-Street bypass, next to Croxdale Autos. You can drive about a hundred yards along this track to the point where it widens out for parking. From here, follow the footpath down the hill through the woods and the place you emerge into open fields beside the river is the downstream end of the beat.

Apart from the very lowest peg, which faces a bend where the river gathers pace and becomes shallower, this section (about quarter of a mile either side of a stream mouth) is fairly deep and slow moving. Either a static or semi-static swimfeeder approach will bear fruit here, with the decent-sized chub in residence often tempted by a variety of baits. It is worth bearing in mind, at this juncture, that this is the part of the Wear where barbel were first introduced and, as such, it might be worth stepping up your tackle strength in the summer as these uncompromising fighters will usually accept most of the more selective baits (luncheon meat, corn, fish pellets, etc.) intended for chub.

For the angler interested in 'smaller' coarse species, there are also lots of dace and the added bonus that Chester Moor has in recent seasons turned up some excellent roach, including fish of over 2 lb. Unlike the barbel, however, the dace and roach will play ball right through the coarse fishing season (16 June to 14 March following, on rivers), with both the stick float and light leger or swimfeeder tactics doing the business, depending on the conditions.

The next upstream section of DCAC bank doesn't occur for another five miles, although there are some interesting stretches before you get there, available by day ticket at Finchale Priory (see directions for Brasside Pond), and free to any angler in possession of a valid rod licence at Ferens Park/The Sands (turn down the hill at the lights on Claypath in Durham City and keep following the road until it comes alongside the riverbank). Both these sections hold considerable reputations for barbel fishing.

DCAC fishing resumes near the centre of Durham City, just upstream of Ferens Park (above Framwellgate Dam). High above, atop the towering banks are the world heritage sites of Durham's castle and cathedral. This stretch is known locally as Prebends, after one of the famous old bridges in the area, and this fine stone footbridge is in fact the one that provides access to the fishery. Parking is available just off Quarryheads Lane, on the opposite side of the river to the cathedral. Follow the path down the hill to find Prebends Bridge.

Prebends is generally regarded as a winter match fishing venue – the whole area being extremely busy with tourists in the summer and university rowers the source of one or two problems in term times. Among the considerable bag weights that this beat provides, specimen dace have become notable addition, including the catch of one recent British Record claimant whose application was refused only on minor technicalities. Most of the other species found in the Wear can also be caught here and while float fishing does work, it is worth noting that some parts of this slow-moving stretch are between 15 and 20 feet deep. Indeed, the depth of the river here is quite an important issue, as right on the point of the bend upstream of Prebends Bridge there is a hole in which barbel can be caught all through the season – one of the few places in the far north of England where this is possible. It's believed that the depth of this particular spot, combined with shelter provided by the high banks on both sides, is the reason the barbel can remain on the feed here all winter long.

Moving along to the next stretch under Durham's control, it's only a hop, skip and a jump upstream to the banks alongside Durham University's Maiden Castle sports complex. My own favourite peg on this stretch is right next to the long jump sandpit and on your approach from downstream – it's just over a half-mile walk through the park (and another stretch of free-water) from the upstream end of Prebends – you can tell you're nearly there by the appearance of a metal footbridge connecting the main sports field with the football pitches over the river.

The spot in question – a large back eddy just upstream of the bridge that revolves around a sandy depression in the riverbed – is fishable at all times of the year by leger or swimfeeder, but this place comes into its own in the winter when the river is up and it's often the only peg for miles around where even a 6-ounce grip lead is going to hold bottom. Chub will provide excellent sport in these conditions and the most effective method for catching them is described in detail in Chapter Fourteen.

Just another five or six hundred yards upstream we come to the next good chub swim – this one unmistakeable and best fished in the evening early on in the season. The river swings to the right as you follow the riverside path from the footbridge and another much larger bridge, carrying the A177 Durham to Stockton road, soon comes into view. The river has been a series of riffles and fast glides in the short distance we've followed it from the last peg, but immediately downstream of the road bridge there is a deep wide pool fed by fast water from underneath the bridge itself.

The trick to this place is to fish in the right spot for the species you want to catch, and getting onto and fishing from the gravel bank below one of the bridge supports would certainly be the best bet if you wanted to run a stick float through the whole swim. While I wouldn't rule out catching larger fish by this technique, there's certainly a far greater chance of your picking up many of the numerous dace and small trout that abound in this fast water. Conversely, if you were to stop about fifty yards before the bridge and look for a steep track down the bank leading to a tight waterside pitch beneath a

The River Wear at Shincliffe Bridge.

tall tree on the sports field side, you'd have found the best place to fish by leger or swimfeeder for the chub.

At this point, the strongest current is mainly coming down the near bank and the hot spot is a deep channel no more than a rod's length straight out from the bank – and stealth is the key. Here, it can pay to dedicate a few minutes to pre-baiting the swim – I use a bait dropper to feed hemp and fish pellets – that literally has to be just dropped in two to three yards out from the bank. The swim can then be fished with either luncheon meat or a pellet hard on the bottom, with either a feeder, or if you prefer, a rolling leger rig trundled through the swim.

Bites can take time to develop here (especially after clunking a bait dropper in five or six times!), but once they do, they're seldom shy. It can also pay to have a change of baits here, too, as because the taking zone is so tight and close to the bank (five yards and you'll have over-cast), the chub will often spook when one of their shoal-mates is hooked. They will eventually come back on the feed but it might save you precious time if you try to coax them back sooner by resuming with a different bait, or even a different technique.

The good chub swims are coming thick and fast on what is now DCAC's Shincliffe stretch, and immediately above the bridge there is another deep sandy pool where the river forms a back eddy similar to the one at Maiden Castle. The main difference here is that this pool is heavily featured on both banks and best fished in the summer months. The fishing rights switch banks above the bridge and access to this peg is by crossing over the river and approaching the opposite bank through the car park of the Rose Tree pub on the Shincliffe side. If you want to bring your car to within reasonable distance of all the fishing spots at Maiden Castle and Shincliffe, there is space for parking on the opposite side of the bridge by the side of the Houghall Grange side road, but the Rose Tree take exception to anglers using their car park unless you intend also to use the pub.

About a hundred yards upstream of the point you leave the pub car park there is another steep track down the bank leading to a sandbank by the waterside, adjacent to the rapids that feed this deep pool. Again, there are several ways in which to fish this spot. One is to feed maggots into the rapids and cast a float towards the deeper water, trundling it down and round into the back eddy. This technique can be made more selective (there are trout, dace, gudgeon and eels that will take maggots here) by substituting the maggot for bread flake and trickling bread feed or free pieces of bread flake down the rapids – an approach more likely to tempt smaller chub. Alternatively, the feeder approach, using either meat or a pellet, will root out the larger chub and even an occasional barbel, although a cast of a good twenty yards into the crease between the main current and circling water is needed to ensure you're fishing in the right place.

Upstream of here, there are one or two more interesting-looking spots, but it's another quarter of a mile to the next really productive peg, past the former railway embankment and on through the next two fields until you reach a sharp bend in the river. The path you follow makes for a rather pleasant countryside walk after you've passed the old bridging site, following the river round until it meets the lane leading from Shincliffe village up to the old hall. The path briefly leaves the waterside as you approach a stand of trees on the near bank and it's at this point that you cut through the long grass on your right to approach the river on the inside of the tight bend.

The correct peg to fish is the one where the backwash from winter floods had formed a small bay on the near side fringed by dense bankside foliage. In a good summer (it changes from year to year), a sand bank will form, giving plenty of space for the angler to get a seat in and stow gear in convenient places. In a bad year you'll need waders, as the water will come right up to bottom of the steep access track down the bank and all your gear will have to stay within easy reaching distance on the bank itself.

There's really only one way to fish this peg: a feeder or bomb cast as hard as you dare into the far bank, where the swift current rushes from rapids just upstream into a deep depression in the riverbed right on the point of the bend. Although the water gets quite deep only a few yards out from the near bank, the killing zone – like downstream of the bridge – is extremely narrow and you will often only get interest in a bait cast right under the foliage that overhangs the far bank. Needless to say, with all the submerged branches and roots, this is a treacherous swim, so expect to lose tackle and be prepared for powerful lunges towards the snags from any of the big chub you can expect to encounter here.

Upstream of here, the river is quite shallow, although there is quite a long run of water adjacent to the stand of tall trees that's well suited to trotting a stick float for dace. After that, the water above Shincliffe Hall has bait restrictions throughout the summer (although bread, which will catch chub, is allowed from 16 June) and being mainly streamy and fairly shallow, it doesn't make for the best winter coarse fishing venue. That does still leave the Ferryham Association water at Croxdale, of course, but again that's a beat better suited to trout fishing, although big chub and the occasional barbel will fall to the usual tactics towards its lower end.

Keeping on the theme of running water, it's now time to move south into North Yorkshire and those hallowed coarse-fishing rivers, the Swale and the Ure. The Swale is often credited with being the fastest-running spate river in England, which may or may not be entirely true, but one thing that certainly is accurate is that if you like your coarse fishing – particularly for barbel and chub – you won't go far wrong on this river.

Unlike many other northern rivers, on which coarse species such as grayling, dace, chub and barbel are gradually found to intermingle with trout the further downstream you go, on the Swale the coarse fishing starts suddenly, just below the Castle Falls in Richmond – a natural obstacle which blocks the upstream passage of every species bar the trout. Thus, within a very short distance of the very beginning of coarse fishing on the river, there are barbel swims of local renown, although it should be noted, as Richmond would be at the upstream limit of the barbel's normal range anyway, that they aren't to be found in residence all the time.

Five hundred yards below the falls, the Mercury Bridge carries the A6136 over the Swale from Richmond town centre. Immediately on the southern side, there's a turning on the left for a public car park accommodating visitors to Richmond's old railway station building, which now houses a variety of local attractions. The local attraction we're interested in is just a few yards in the opposite direction – a deep pool in the river set into a sharp bend with a high bank and cliffs on the far side. Richmond and District Angling Society control the fishing rights in this area, and although membership requires the applicant to live within 5 miles of Richmond, day tickets for over 4 miles of coarse fishing between Richmond and the village of Great Langton are available for just £6 from the Tourist Information Centre in Richmond's Market Place.

Downstream of Richmond town, there are lots of shallow pools and runs that would be of interest to the dace or grayling angler, but the next place of importance to the barbel or chub fisherman is 2 miles downstream near the village of Easby. Easby is found by driving out of Richmond towards Scotch Corner, then after a set of traffic lights, bearing right down the B6271 towards Brompton-on-Swale. About a mile down this road there is turning on your right signed for Easby. Follow this road for about half a mile before turning right again in the hamlet of Easby and going downhill to a small car park intended for the abbey ruins on the right.

Once parked up, instead of heading towards the abbey, walk instead down the lane heading in the opposite direction, which eventually comes right to the top of a high bank beside the river before heading through a small wood and on towards an old railway bridge. Cross the bridge and immediately on the other side scramble down the embankment into the field that borders the river on the opposite side to the car park. Follow the fence alongside the river down to the corner of the field and right in front of you there should be a big pool, with the near bank forming the gravel strewn inside of a bend.

This place is known locally as Easby Bend and when barbel are in the area it is often the best place to fish for them. The problem you now have is where to cast, as the pool is both wide and deep (most of the way across), and both the barbel and chub have a reputation for positioning themselves in different places on any given day. The only way to find out where is by experimenting and while pre-baiting does help, the sheer size of the pool means you're only ever going to give the fish a taste of what's to come – rather than hold a group of them in any given spot.

Leger or swimfeeder is always going to be the most effective method for fishing for chub and barbel, with strong tackle essential, especially when bearing in mind the snaggy nature of this rock-strewn riverbed. A tip, when fishing here, is to use a weaker length of nylon to connect the weight or swimfeeder, so if that's what gets stuck under a big rock, at least you'll get the rest of the rig back. Also, never drag the rig back on a tight line as some anglers do after casting – instead keep the line slack until the rod is in the rest and even then keep pressure on the weight to a minimum. This might be all well and good when casting right in under the far bank across the main flow, but anything more than a slight bend in the rod tip is likely to result in a snag.

Like for the Wear, the Swale's chub and barbel should respond best to luncheon meat or a pellet, with hemp and/or small pellets in the feeder. Caster, when you could get it, used to be a very effective bait for the bigger fish at Easby and had the added advantage that if you caught small trout, grayling and dace to begin with, then nothing, you knew that chub or barbel had probably moved into the swim and would likely be up next. Likewise, sweetcorn is another old favourite that could turn up the odd chub or barbel on a day when others were ignored. Float fishing is also productive on the bend at Easby (and is very effective for catching the grayling in the rapids below it) but like further north this will usually only account for the smaller species in the swim.

Half a mile downstream from Easby, Red House Farm, on the opposite bank, is another well-regarded spot for chub and barbel fishing – a deepening pool below a wide set of rapids – although personally I've never fared that well there. There are other good spots just downstream at Broken Brae and Brompton-on-Swale, but it's another 12 miles downriver before we reach the next really good spot under Richmond's control.

If, instead of turning off for Easby, you continue on the B6271 through Brompton-on-Swale and bear right at Scorton, you'll eventually arrive at the village of Great Langton. Just before you enter the village, there is a minor road turning on the right that immediately crosses over the river on a single track metal bridge. Park in the lay-by beside the B6271 on the Langton side and cross the bridge. On your left, downstream of the bridge, is water controlled by the Kirkby Fleetham & District Angling Association, which is fly-only for trout. On the right, on the opposite bank to Great Langton, is the last downstream section of Richmond water and the deep hole close to the Langton bank about 20 yards above the bridge is an excellent place to fish for both chub and barbel.

Great Langton is far enough down the 'upper Swale' for it to be a more reliable barbel fishing mark than the places nearer Richmond town. Even so, you'll still get days when the barbel are nowhere to be found, and in any case chub will ordinarily outnumber their cousins by at least five to one. The river is much narrower here than at Easby, despite our being over 10 miles downstream, and with the killing zone generally restricted to the channel down the far bank (and on bright days, right under the far bank), baiting up with a 'dropper' is generally worthwhile.

Once you've baited up, what goes for Easby usually applies at Great Langton and, like at Easby, you can also float fish, although this is less likely to get the bigger species. Along with many trout, the Swale here also contains dace and grayling, with the occasional perch showing up from time to time. All in all, the water downstream of Richmond comprises an excellent mixed coarse fishery (and if you pack a fly rod, it's also an excellent trout fishery as well). With all the spots to be fished, £6 for a day is rarely going to be money wasted.

About twenty minutes further down the A1(M) from Scotch Corner (the turn-off for Richmond) is the Dishforth Interchange where the Great North Road meets the A168 from Thirsk. If you bear left here and follow the signs that direct you onto the A168 towards Thirsk, after about a mile the first exit takes you up a slip road. At the top, take the right turn signed for Cundall. There are two very good day-ticket fisheries down this road and the first, Cundall Lodge Farm, comprises about a mile of fishing in fairly slow, deep

The author with a 9¼-lb River Swale barbel.

water that fishes well for barbel, chub, perch, pike and bream. Tickets (£5) are available at the farmhouse, which is by the main road just before the village of Cundall. Cars can be taken down and parked on the bank – turn in through the gate that is just before the house.

The second day-ticket fishery is just a couple of miles further downstream on the opposite bank and is made up of more streamy water controlled by the Helperby & Brafferton Angling Club. To find this water, continue on through Cundall and after another couple of miles you'll cross the river at Thornton Bridge. On the opposite side of the bridge, take the first left turning and the first left after that, then follow this narrow lane for about a mile until you approach some farm buildings on the right (Fawdington). Day tickets (again £5) are available from the farmhouse, or alternatively from the village shop in Helperby (continue straight on for another mile after Thornton Bridge). This water fishes well for most coarse species but is particularly good for barbel and pike.

Finally, it would be remiss of me not to mention the beautiful River Ure at Ripon. The Ripon Piscatorial Association controls around six miles of double bank fishing on the Ure up and downstream of the small city of Ripon. Membership details can be found on the association's website (www.ripon-piscatorial.co.uk) but £6 day tickets for the river can be bought from Bondgate post office, Ripon News on North Street and Ure Bank caravan site (all in Ripon) as well as Fish 'n' Things Tackle Shop, Horsefair, Boroughbridge.

My own favourite spots on the river are all just up and downstream of North Bridge, which is the bridge you cross if you take the right turn at the first roundabout on the A61, coming in from the direction of the A1. Going downstream (park in All Hallowgate, the first left turn after North Bridge), go through the gate on the left into the field alongside All Hallowgate. Follow the footpath under the new bridge carrying the A61 Ripon bypass and along the right bank of the river for about quarter of a mile. The place you are looking for is a right-hand bend in the river, where the far bank rises high above the surrounding land.

The bend is a set of rapids and on its inside there is a stony bank. Go almost to the bottom of the stony section until you are facing slightly deeper water on the far bank. Fishing luncheon meat on a swimfeeder or leger rig in close to the far bank will catch chub here and a stick float with maggot trotted through from near the top of the bend will take dace and the occasional chub.

Upstream of the rapids there is a deeper pool which can be fished from a peg set between trees about 30 yards above the stony section of bank. A swimfeeder with luncheon meat or caster thrown across to the far bank will stand a good chance of catching a barbel or chub, while float-fished maggot will often tempt good-sized perch and dace.

Upstream of North Bridge, the fishing rights are on the opposite bank, so rather than crossing North Bridge (at the same turning off the A61), continue straight across the mini-roundabout before the bridge and carry on past the houses of Ure Bank. Shortly after the houses end there is a car park between a pond and the river. If you cross the stile and follow the river upstream there are several good places for fishing, with chub, barbel and dace the target species. You could also choose to adopt a more roving approach here, as the RPA's fishing rights continue for about another mile upriver. A float-fishing set-up using a maggot or worm for bait will account for grayling and dace (as well as trout) in some of the stream-like spots, and might possibly tempt even the odd chub or barbel.

CHAPTER NINE

Coastal Roots: The History of Sea Fishing from the Coast of North East England

By now you could almost be forgiven for thinking I'd forgotten about that magnificent coastline of Northumberland and Durham and the superb sea fishing it offers. Far from it! For many years, alongside the River Wansbeck at Morpeth and Bolam Lake, the coast and the two estuaries between Newbiggin and Blyth in Northumberland was a focal point for my fishing attentions. For thousands of others, the whole North East coast comprises the fulcrum of their angling endeavours; and for all who similarly aspire, these next chapters are for you.

The traditions of angling along the coast of North East England are almost as deeply ingrained as those of the now dwindling coastal net fishing industry. The ports and harbours from Tweedmouth, in the north, down to Whitby, roll off the tongue like a who's who of shore fishing marks: Seahouses, Craster, Amble, Tynemouth and Shields, Hendon, Seaham, Hartlepool. In stark contrast to the inland waters, fish stocks in the North Sea are now sadly in decline but the anglers remain and fish are still there for the dedicated sea fisherman prepared to hone his skills.

Back in the 1970s and 1980s, the problems of overfishing were only just beginning to be realised. In Frank Johnson's *North East Angling Guide*, Dave Higgins, Press Secretary of the Northern Federation of Sea Angling Societies asked, 'What has the North East coast got to offer the shore angler? A very great deal, as I'm sure you will agree if you know the area.' Indeed there was something for everyone back then – from the specialist who could cast 200 yards to us kids mucking about with flatties and small pollack.

'The coastline is very varied,' Higgins wrote, 'offering tremendous contrast, even in areas close to each other. There are low-lying areas with long and really magnificent sandy beaches flanked with sand dunes, with some of these areas, mostly in Northumberland, interspersed with rocky outcrops. These are contrasted by fairly steep sloping and unattractive coal-strewn beaches. Then there are the more rugged stretches of cliffs and rocks, with kelp-lined gullies and small sandy bays and adding further to the wide character of the North

The beach at Cambois, looking south towards the mouth of the River Blyth.

East coast, there are estuaries and harbours, where the shore angler can enjoy his sport on the grimy jetties, where angling is permitted, well up the main rivers of the Blyth, Tyne and Wear.'

The range of species that could be caught from the shore in North East England was never as varied as in other parts of the country, but it was the size of some of the fish that got pulses racing. 'The main quarry is the codling,' continued Dave Higgins, 'which are taken roughly throughout the year, although the main season runs from early September through to January, February or March, depending on the particular season and, to some degree, the part of the region.'

Coalfish and whiting were the other main winter targets, with dabs, flounders, pouting and plaice the summer staples; however, it was the mackerel that provided the real attraction when the warmest weather arrived. 'Mackerel are either loved or hated by sea anglers,' Higgins mused. 'There isn't any half way. But whether you're interested or not in mackerel fishing, the season is still a very short one. It normally extends from early or mid-July to late August or early September and unlike the pouting, where the average size has fallen, the mackerel have increased in size over the last few years. Ten years or so ago, you could almost be sure their size would be in the very narrow limits of 15 oz to 1 lb 2 oz, but recently, specimens of 1 lb 8 oz have been common, a few even topping 2½ lb.'

Jack Charlton remembered his forays for mackerel with the same fondness that had given him that lifelong love of the trout: 'We'd go off to the rocks at Newbiggin maybe three or four evenings a week after school in the summer. There might be anything up to a hundred people stood waiting for the mackerel to show, when they chased the sprats inshore. Suddenly the water would start to boil and a hundred voices would yell excitedly, "They're in, they're in!" That would be the signal for frenzied activity along the shore as hundreds of lines hit the water.'

Jack's mackerel set-up was primitive: 'We used a fly made up of silver paper and a piece of goose feather, and what was known then as a Scarborough reel. It was a long way removed from the modern gear, which can throw a bait for miles, but once you got the hang of it, you could cast out fifty yards or more, reel in, and if you got lucky, you might have three or four mackerel every time.'

And Jack didn't confine his sea fishing to just the summer months: 'In the autumn, we would go and fish the storm beaches. This was a pursuit for only the most hardy. We'd set off after school and, providing it didn't rain, we'd stay there until the dawn broke and then walk the three or four miles back home to be ready for school. God, we must have looked a motley crew. I mean we'd have two or three pairs of socks on, two pairs of trousers and a couple of jumpers – anything just to keep you warm against the chill of the night. There might be fifty or sixty of us there a night, young and old, searching out the codling which came in on the surf. I never got anything enormous, once or twice a six or seven pounder, but I could never wait to get home and show my catch off to my mother.'

Of course, as Dave Higgins alluded to, not all the saltwater angling in the region was confined to the open sea. As we already know, the North East is home to a great many mighty rivers and there are no fewer than seven major estuaries of varying fishing potential from the relatively small, like the Aln at Alnmouth, to the Tyne, whose tidal reach goes some twenty miles inland. To my mates and me, back in the early 1980s, two of these estuaries in particular became our main venues of choice, on those days we chose to give the trout and perch some peace and quiet.

My home river, the Wansbeck, flowed to the sea at a picturesque spot known locally as Sandy Bay, forming a quintessentially shallow, muddy estuary that could almost have been mistaken for the iconic River Hamble on the south coast; such were the multitude of pleasure craft that moored in its tranquil waters. Near the point at which it met the sea, the estuary narrowed at a neck in the river formed by a coastal spit and, at low tide, a deep hole adjacent to a mooring point for boats could be fished for flounders, or 'flatties' as we called them.

Tackle would be simple – a beaten-up old leger rod, no longer considered of use for freshwater angling, and line of about 5-lb breaking strain – as the fish were rarely bigger than a pound or so, and more often than not smaller. But this shallow inlet was home to thousands of flatties (and also, back then, to vast numbers of mussel beds that provided easily accessible bait), so catches could regularly run into the hundreds between even a small group of anglers in the long days we used to spend there. Even the relentless chill of the cold wind that blew in off the North Sea could be counteracted, by fires built in the sand that were made of driftwood and sea coal – there was an answer to every problem at this mark!

Only 2 miles down the coast, the estuary of the River Blyth was of a completely different order. In those days, this was a working river, more like the Tyne in character, well over 30 feet deep in most places and half a mile across at its widest point. In the 1970s and 1980s, the North East was still one of the most important coal-producing areas in Europe and the Blyth, with its deep sheltered estuary, was a major port in the trade.

Among the most iconic symbols of coal export in the North East were the coaling staithes, structures that existed on all the major estuaries, sadly almost all now gone. These were gargantuan wooden jetties that towered above the water, half a mile in length in some cases, and from which coal could be transferred directly, by means of gravity, from the coal trains stationed above into the holds of the waiting ships moored below. The Blyth had many of these structures, and given the depth of the riverbed into which their timber piles had been sunk, they could be prodigious marks when fished from the gangway situated just above the high-water mark. In winter, the rough seas pounding the rocky shoreline adjacent to the mouth of the River Blyth would frequently result in sizeable pollack and coalfish being swept into the estuary on a flooding tide – sometimes, even codling. It was always rough going, fishing quarter of a mile out into the river, but rarely unyielding.

Sadly, unlike almost all the other types of angling practised in the area, the intervening decades have not been kind to sea fishing in the North East of England. Late twentieth-century industry – in the guise of warm water outfall from the power stations at Blyth and Lynemouth – did briefly bring such 'exotic' species as bass and mullet close to the Northumberland coast (indeed global warming has occasionally beckoned them north in very warm summers too), but like in all parts of the North Sea, the spectre of commercial fishing looms large. The last quarter of a century has seen a marked decline in the quantities of sea fish caught from the North East coast, and where bass and mullet have been seen for the first time in relatively small numbers, these cannot make up for the shortfall in species formerly common. One can only hope that, given the increasing restrictions being brought to bear on the net fishing industry, benefits similar to those seen in the populations of returning salmon and sea trout can be mirrored in non-migratory sea fish populations. Unfortunately, so far things don't look all that promising.

CHAPTER TEN

The Court of Old King Coal: A Guide to the Best Sea Fishing Marks along the Coastline of Northumberland and Durham

It goes without saying that the nature of rod fishing from the beaches, rocks, and piers of North East England has changed beyond recognition since the 1970s and early 1980s. Quite apart from the declining fish stocks, I hardly recognise some of the places I used to fish whenever I go back. Yet, the coastline itself has actually changed little, save for the odd bit of erosion here and there. It's just that the paraphernalia that used to go with it, the spectre of indelible images like Blyth Power Station, can leave a ghostly feeling as if something just isn't right.

For sure, certain marks such as coaling staithes, places we used to take for granted up and down our coast and estuaries, have vanished en-masse, but most of the others still remain. The list of fishing marks along the Northumberland and Durham coast is still lengthy and provides a rich variety of fishing opportunities in respect to both the locations themselves and the time of year.

We'll start in the area I know best and firstly visit Newbiggin, a fishing village located at the coastal end of the A197, a road that runs from Morpeth via Ashington and intersects the A189 Spine Road near its northern terminus. Newbiggin can fairly be divided into three marks, with those straddling the sandy seafront the ones most non-anglers will probably recognise.

The first two marks are rocky in nature and comprise the headlands at either end of the beach. Both will fish well all year round for pollack (generally small) and the occasional codling, but come into their own during the warmer summer months as fine locations for catching mackerel on spinning gear. For Church Point, follow the Front Street right to the top (turn left at the end of the A197) and, at the turning point for buses, take the track leading straight ahead towards the church. For the other mark, Spital Point, you'll need to turn right at the end of the A197 and find a parking space in the village about half a mile along the B1334. Spital Point is found by following one of the paths down to the southernmost end of the promenade and turning right towards the rocks.

The third Newbiggin mark is a storm beach that lies between Spital Point and the mouth of the River Wansbeck about a mile further south. This mark is sometimes referred to as Sandy Bay and lies at the foot of cliffs upon which stands a well-known caravan site of the same name. This is still an excellent beach for catching cod in the winter, while pollack, whiting and flatfish will feature in the summer months, with gulleys running along the sandy foreshore that can be spotted at low tide. Locals tend to prefer lugworm or crab as bait for the cod, although ragworm and fish-baits will catch as well, particularly the smaller species.

It's also worth noting that the poisonous lesser weever (known locally as the 'granny fish') will sometimes figure among catches here, so don't handle any small fish that look in any way suspicious. Sandy Bay fishes best at high tide and is accessed by either following the directions for Spital Point and continuing on south past the rocks, or by finding a parking space near to the Sandy Bay Caravan Park, whose turn off is on the Newbiggin side of North Seaton Roundabout (A189/B1334) on the B1334.

Ignoring the estuary of the River Wansbeck (which will fish but, as we already know, yields mostly small flatties in large numbers), immediately to the south is the 3-mile expanse of Cambois Beach, which starts as a gently sloping, clean and sandy beach (at the Wansbeck end) but becomes increasingly rough the closer to the mouth of the River Blyth you get. At its northern end, Cambois will still fish as a classic winter cod beach; although being cleaner, it lacks the status of its more northerly neighbour. During the summer, a catch from here will usually feature mainly flatties at high tide (flounders predominate in the Wansbeck itself), with ragworm and fish-baits usually doing the trick, although the occasional bass can't be ruled out.

At its southern end – beyond where the hot water outflow from Blyth Power Station used to be – Cambois beach becomes progressively rough and then rocky, and is essentially a winter or a night-time summer mark for specialist cod anglers, when, as the local saying goes, 'there's a bit of a sea running'. Beyond here, and past the three kiln-shaped silos of Alcan's bauxite refinery, Cambois beach becomes a narrow peninsula, sandwiched between the sea and the last half mile of the River Blyth estuary. Near the point, there are a couple of old landing stages that can be used for fishing at high tide into the mouth of the River Blyth, but neither can be considered prime spots, nor the long-dilapidated north pier which nowadays houses eight large wind turbines.

The whole of Cambois beach can be accessed from the A196 at Stakeford (on the opposite side of the River Wansbeck to Ashington) by taking the A1147 towards Bedlington Station and shortly afterwards turning left down the road to Cambois. After two miles, take a left turn beside some old school buildings, pass under the A189 flyover and reach the sea front at the first part of the scattered community of Cambois. Shortly after reaching the sea front,

Wide and deep, the estuary of the River Blyth is a popular mark all year round and the scene of the famous closing sequence in *Get Carter*.

there is a car park on the left for the northern end of Cambois beach. For the southern end, continue on into the main part of Cambois village (about a mile) and park considerately. Find the path that starts near the mini-roundabout at the centre of the village and follow it underneath the railway. This brings you out onto the southern end of Cambois beach, which continues for almost two miles south to the mouth of the River Blyth.

Following these directions will also bring you to one of the last remaining staithes on the River Blyth estuary (or what's left of it, anyway). Continue on past the mini-roundabout to the small settlement of North Blyth – the old ferry point from the Cambois side over to the main town. Parking is possible beside the former landing stage on waste ground next to the entrance to Battleship Wharf.

If you've ever seen the iconic film *Get Carter*, this is actually what remains of the structure that Michael Caine (Carter) climbed onto at the beginning of the closing chase sequence. The railway gantries that once towered above the remaining timberwork have long since been dismantled and what's left is now essentially a jetty stretching for about a quarter of a mile alongside the north bank of the estuary, with deep water immediately out from its walkway.

In summer months, fishing off this structure will produce lots of smaller fish, although bigger flatfish are possible if you can locate them. Autumn and winter will see the bigger species entering the estuary, with either a good cast out into the middle or a bait dropped in at the side being effective. There are even lights on the staithes, so the mark can be fished into the hours of darkness, with autumn in particular being a productive time for catching codling, and larger pollack and whiting, close into the timber piles.

Lastly, only just within the Blyth estuary, is Blyth Pier, which marks the southern extent of the river mouth. Unlike many other similar constructions in and around Blyth, this is actually a real pier and highly regarded as a spinning mark in summer, with mackerel and coalfish the target species. Casting out towards the south beach side is generally best, as the mouth of the river can be quite snaggy, with the end of the pier, past the bend, most favoured. Blyth Pier is found by leaving the A189 Spine Road about three miles north of where it crosses the A19 and taking the A1061 towards Blyth. Upon reaching the A193 coast road roundabout, continue straight ahead towards South Beach and after half a mile turn right into the Port of Blyth and carry on another quarter of a mile to find the pier.

There are, of course, very many highly productive shore marks along the Northumberland coast, with Newbiggin to Blyth comprising only about 8 miles of shore and estuary out of a total of 60 miles from Berwick to Tynemouth. Some marks are well known and others closely guarded secrets known to only a small number of highly dedicated specialist anglers. Here are just a few of the better known ones, with directions and a brief description.

Embleton Bay is a fairly clean two-mile-long beach situated between Newton Point and the imposing cliffs and castle of Dunstanburgh to the south. It fishes particularly well at low tide and the target species are mainly flatfish, with the best baits lug or ragworm. For Embleton, follow the A1068 north as far as Lesbury (about a mile north of the turn off for Alnmouth). At Lesbury, turn right onto the B1339, continuing to follow the brown 'Coastal Route' direction signs. Carry on following both route signs until, just after the B1339 becomes the B1340, you take a right turn for High Newton-by-the-Sea. Continue through High Newton and, where the road swings round to the right, continue on for another mile over the brow of a hill until the sea and Embleton Bay come into view. There is a car park on the right and the beach is just a short walk away.

Craster is only a few miles south of Embleton Bay and offers several excellent rock marks, situated either side of this tiny fishing port, which can be fished at either high or low tide. Craster is found by turning right off the B1339 about three miles north of Longhoughton and following a minor road down to a car park on the edge of the village. The rock marks are situated on either side of Craster and include Cullernose Point to the south. Alternative parking for other rock marks further south of Cullernose can be found by taking a right turn at the crossroads before you reach Craster and heading south towards Howick.

About seven miles south of Craster is another clean, sandy beach-casting mark that lies in the shadow of another famous Northumbrian castle. Going north, Warkworth beach is found by following the A1068 through Warkworth and then going right at the first turning after crossing the River Coquet (or left at the last turning before the Coquet, if you're coming from the Alnwick direction). Follow the lane down to a car park and the beach is a short walk over the dunes. Like Embleton, fishing lug or rag will produce mostly flatties with the occasional bass in high summer, although this beach fishes best an hour or two either side of high tide.

Like Sandy Bay – and the Wansbeck – the southern end of Warkworth beach is bounded by the mouth of the River Coquet, although there is a small breakwater that can fish well for mackerel with a spinner in the warmer months. On the opposite side of the river, the small but bustling port of Amble can get quite busy during the holiday season. Outside of the holidays, its recently renovated riverside jetty makes for an excellent fishing platform from April through to September, although you'll be hard pressed to catch anything besides flatties in the estuary itself. Amble is a mile south of Warkworth on the A1068. Coming north, turn right at the roundabout just as you reach the outskirts and follow the signs down to the harbour to park. The riverside jetty begins on the quayside and, generally speaking, the nearer the sea you fish, the more bites you'll get. And watch out for those seals that like to swim across from nearby Coquet Island!

From Amble south to Newbiggin, the coastline is widely varied. Immediately south of the Coquet is a rocky headland beside Low Hauxley which almost immediately becomes the 6-mile sandy, dune-fringed expanse of Druridge Bay. South of Druridge the small village of Cresswell sits atop another rocky headland, with kelp-ridden sandy inlets just to the south – the nearby feature at Snab Point being a favourite mark for local anglers. Druridge, Cresswell and Snab Point can all be accessed by taking the turn for Druridge at the large roundabout on the A1068 at Widdrington. This becomes the coast road beside the National Trust car park for Druridge Bay.

South of Blyth, there is another 3-mile expanse of sandy beach that is often busy in the summertime, before another stretch of rocky shoreline commences at Seaton Sluice. South of here, Whitley Bay, Cullercoats and Tynemouth are mixture of beach and rocks, with several marks that are popular with anglers, although, like at Amble, all will be popular with holidaymakers during the summer. The southern end of the Northumberland coast is marked by the wide mouth of the River Tyne, although the mile-long stone-built North Pier is unfortunately not open to anglers. All marks from Blyth to Tynemouth are close to the A193 coast road.

Over the river, South Shields is the northernmost town on the County Durham coast and, like the places immediately to its north, this too is a popular resort with tourists in the summer. Fortunately for the angler, most of them frequent the parks, beach and funfair about half a mile south of the mouth of the Tyne, leaving a couple of excellent fishing marks that are accessible relatively easily throughout the year.

Unlike its northern counterpart, the South Pier at South Shields is open to anglers and this mark is amenable to all forms of shore fishing, winter and summer, although the ground on its inner side can be doggedly snaggy. Warm weather in July and August can often bring large numbers of mackerel into the mouth of the Tyne and around halfway along the South Pier there are several good spots for spinning. Also dissimilar to the North Pier, this giant breakwater juts out into the sea a good way south of the actual point the estuary meets the coastline. This results in a short stretch of beach, the North Herd Sands, which is actually within the mouth of the river and this less fashionable stretch, too sheltered for surfers and the like, will fish for flatties at high tide – although a longish cast too close to the South Pier may locate the rocks that are never that far out.

At the northern end of Herd Sands, there is another much smaller stone pier structure called the Groyne, a mark which comes into its own in late autumn for decent-sized codling, whiting and coalfish. This place fishes best during the two-hour period either side of high tide, with a pennel-rig often accounting for the better cod.

To get to this part of South Shields, follow the A194 all the way into town and at the sixth roundabout after Tyne Dock (the last before you go under the metro line) take the first left turn onto the B1303. Follow this road round as it comes alongside the south bank of the river and after a couple of miles and a gradual 180-degree turn, the Groyne will appear on your left, with the South Pier ahead. There is a car park on the left-hand side of the road adjoining North Herd Sands.

Beyond South Shields there are the precipitous heights of Trow Rocks, which adjoin the southern end of Herd Sands, near the Gypsies Green Stadium. This is a popular spot for fishing, despite the risks, although anyone not familiar with the area would definitely be best advised to think twice, as anglers have fallen to their deaths here in the past. South of Trow Rocks, the coastline remains almost continuously rugged as it goes around the headland of Souter Point (home to a distinctive lighthouse, as well as rifle ranges) and down as far as Whitburn, by which time we are already six miles south of the Tyne.

South of Whitburn, the next seven miles of coastline are occupied by the large town of Sunderland – which itself has several favoured marks either side of and within the mouth of the River Wear. The North Pier at Sunderland, guarding the northern entrance to the river, is a popular and easily accessed mark that fishes best at high tide. Worm baits, mackerel, crab, mussel and feathers will all take fish, with a variety of species being caught here, although

Shore fishing on Warkworth beach.

it can be snaggy close in to the pier wall – the closer to the end of the pier you can fish, the better.

Not far up the River Wear itself, staying on the same side as the North Pier, are two very popular deep-water estuary marks. The Glass Centre is best fished from low water until high tide with a cast of 40 yards into deeper water required to locate the better fish. Codling and whiting are the target species, with the mark at its best in the colder months.

Rat House Corner is just a few hundred yards upriver from the Glass Centre and, in addition cod and whiting, a 40–50-yard cast into the river will take flounders and coalfish, with the best tactic for both these river marks a two-hook flapper rig.

To get to the North Pier, the Glass Centre and Rat House, take the A183 towards Roker from the north end of Monkwearmouth Bridge (A1018). The National Glass Centre is on the right-hand side, going towards the coast, only a few hundred yards after turning onto Dame Dorothy Street (A183). For the North Pier continue along this main road as it becomes Harbour View and on past Sunderland Marina. Approximately one mile on from the Glass Centre, turn right into Pier View at the mini-roundabout as you reach the seafront. Follow this road as it swings round to the left and park next to the seafront. The entrance to the pier is to the left, looking from the end of Pier View.

To the south of the River Wear, Hendon Prom is an open sea mark favoured by locals from which cod, whiting, bass, flounders and coalfish can all be caught on a variety of baits, with the best time for fishing being the autumn and winter months. Take the A1018 Commercial Road north from Ryhope, or south from Sunderland Centre, and turn towards the seafront just after (from Ryhope) or before (from the town centre) three gasometers by the roadside. Drive down underneath the railway and the road swings round to the right to become the promenade. You can park where you choose to fish.

Three miles further south from Sunderland lies the much smaller port and former mining town of Seaham. There are several good beach marks around Seaham, all renowned as winter storm beaches for cod fishing, with the Blast and Dawdon beaches situated just to the south of the town and Hall beach to the north. Within Seaham itself (and just to the south of Hall beach) there is also Seaham Prom, a mark that can fish well either side of high tide for cod, as well as whiting, coalfish, flatties and, in high summer, bass. The favoured baits here are lugworm or crab for cod, while the other species will take lug, rag or fish-baits.

For Seaham, going northbound, leave the A19 at the second exit signposted for the A182 (N.B. Not the turn off for Houghton-le-Spring) and follow the relief road round into the town. From the north, leave at the exit for the B1404, turn left and follow signs down into the town.

South of Seaham, the County Durham coastline continues for the next twelve miles as a mixture of sheer magnesium limestone cliffs and beaches, with Horden Beach (turn off the A19 at Peterlee and follow the B1320 down towards the coast until you reach the old pit village of Horden) just one of those held in local esteem. The southernmost town on the Durham coast is Hartlepool and here we find the last marks worthy of mention on our journey.

The favoured fishing marks around Hartlepool are both piers, which give the angler whose casting ability falls short on the beach a better chance of reaching deeper water close to the bigger fish. The northernmost is Steetley Pier, situated just to the north of the old town next to a redundant chemical works, and both the pier itself and the old works chimney beside it are clearly visible as you approach from the north along the A1086 coast road. Decent-sized codling and whiting are the main targets here in winter, with flatties, mackerel and bass all showing in warmer weather – pennel tackle being the advised rig of choice, together with lugworm, squid or razorfish for bait. To get there take either the A1086 or the A179 (coming off the A19 from the north or south) and at the roundabout where they meet take the A1049 down towards the sea front. It is best to park close to houses or shops here and walk down under the railway to the pier.

Middleton Pier is closer to the centre of Hartlepool near the town's marina and can be found by taking either the A689 from the south or the A179 from the north (both meet the A19 nearby). Where the two roads join at a roundabout, go down Middleton Road in the direction of the harbour. Middleton Pier is in front of you (unfortunately there is a gate at the entrance to the pier which can sometimes be locked) and it fishes well all year round and at all states of the tide, with codling, coalfish, whiting and flatties showing, along with mackerel in the summer. The same baits that catch fish at Steetley Pier will also work here.

Going south from Hartlepool, there is a further three miles of mainly sandy beach before Tees Mouth is reached at the North Gare Breakwater. This marks the southern boundary of the Northumbrian coastline, as south of the Tees the rugged coastline of North Yorkshire veers off on a south-easterly tangent towards Flamborough Head. But what lies to the north should be more than enough to keep you occupied!

PART TWO

North East Fishing Seasons

Official Start and End Dates for the various Rod Fishing Seasons in Northumbria & Yorkshire

Brown Trout*	(Northumbria)	*Start* 22 March	*End* 30 September
	(Yorkshire)	*Start* 25 March	*End* 30 September
Migratory (Sea) Trout		*Start* 3 April	*End* 31 October
Salmon**	(Northumbria)	*Start* 1 February	*End* 31 October
	(Yorkshire)	*Start* 6 April	*End* 31 October
	(River Tweed & tributaries)	*Start* 1 February	*End* 30 November
Coarse Fish**		*Start* 16 June	*End* 14 March (following)

Exceptions and Dispensations:

*The season for Brown Trout on Kielder Water, Broomlee Lough, Crag Lough, Greenlee Lough, Derwent Reservoir, and East & West Hallington Reservoirs is from 1 May to 31 October.

**All Salmon caught before 16 June in Northumbria and Yorkshire must be returned to the water immediately with as little injury as possible. All salmon caught before 30 June from the River Tweed and the tributaries thereof must be returned to the water.

***There is no statutory close season for coarse fish on any stillwaters in the region, however individual owners may close their fishery if they wish. ALWAYS check before you fish.

There is no close season for rainbow trout in any stillwaters.

CHAPTER ELEVEN

Spring: Nymph, Wet and Dry Fly Fishing for River Trout

There are few more pleasant thoughts when you're caught in the grip of a cold and unrelenting winter than that of walking along a riverbank, fly rod in hand, on a warm and sunlit spring evening. Of course it doesn't take me to tell you that it's never quite that straightforward up here in the North East of England, given the weather that usually greets the opening day of the trout season on 22 March. Quite apart from the threat of a late winter 'sting in the tail' consisting of hail or even snow, the cool, breezy weather more normally associated with a typical North East spring day can in itself make trout fishing at this time of year a less than comfortable experience.

Even the occasional bright, mild day, jammed incongruously into a spell of generally colder weather, can sometimes give the wrong impression and lead the attentive angler to conclude that the fact he can see no fish activity on his chosen stretch of river means there are no trout worth fishing for. More often than not, however, despite the fact that the sepia-coloured water can look deceptively devoid of life, this is only because the first spring olives (small and medium sized up-winged flies of the order *Ephemeroptera*), for whose emergence the trout are patiently waiting, have yet to hatch due to low water temperatures. Until they do, the trout will see little point in expending the extravagant bursts of energy we will see soon enough – but only once the weather eventually warms up. Until it does, this cold-blooded creature will remain comparatively docile, ever watchful, in readiness for the moment its instincts will tell it to start chasing the increasing number of newly emerging insects.

Spring trout fishing with the fly is more often than not, therefore, a waiting game – or alternatively it could be a time to try out the upstream nymph technique that is the staple of the autumn grayling angler (flick ahead to Chapter Thirteen for a detailed description) – for in skilled hands this method can be equally as effective in the early weeks of the trout season. For those less confident about casting a fly upstream in the often swift river-flows of early season, there is also the option to cast 'across-and-down', in the manner of the

The River Wansbeck at Morpeth. The swift water in the right of this picture feeds a pool under the left bank that will give trout to a well-presented fly.

classic wet fly technique. Trout will readily grab at a deeply fished weighted nymph swinging across a brisk current (see Diagram 1 for the general outline) – just make sure your knots are well tied as takes can be savage! And here's a tip, if the river's still freezing cold with no sign of activity: try a brass- or tungsten-beaded Pheasant Tail Nymph or Gold Ribbed Hare's Ear as a point fly in combination with a traditional early season wet fly pattern, such as the March Brown, on a dropper. Sometimes the added depth the weighted nymphs enable you to fish in a fuller current will put the one thing a big early season trout just can't resist right on his nose.

Needless to say, the very concept of fly casting, with its inevitable complexities, can be an instant put-off to many anglers more familiar with other techniques – and that's before we even consider the need to understand such unique concepts as fly line ratings. To be fair, the need for professional coaching is probably more appropriate to fly fishing than for any other angling discipline, yet this is really only so for fishing on stillwaters. Because of the shorter distances involved, technique is not generally so critical to casting a fly on a river, and there are other factors – most of which are outlined in this chapter – that are of far greater importance. The use of a correctly balanced rod and fly line that is well suited to your chosen river will usually mean that a bit of practice is all that's required to get you on your way.

We can assume that the weather will warm up eventually, as the days get ever longer through April and May. By now the olive hatches should be increasing and the brown trout will spend much of their time higher up in the water,

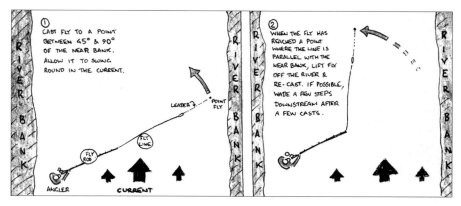

Diagram 1: Wet fly fishing in the 'across and down' style.

attempting to capture nymphs on the final part of their ascent to the surface to hatch into their adult form. They'll also target those insects in the throes of metamorphosis as they struggle in the surface film itself, trying to rid themselves of their nymphal shuck. It follows, therefore, that while dry fly fishing – in which the adult form of an insect is imitated – becomes increasingly successful the longer the season goes on, those techniques which mimic the ascent and emergence stages of a hatch can also be devastatingly effective, provided you choose a close enough imitation of the fly on which the trout are feeding. The best way of achieving this is by the classic wet fly technique, which can employ either orthodox wet or (un-weighted) nymph fly-patterns, and on typically sized North East rivers requires a 7–9-foot fly rod, corresponding floating fly line and a seven to nine foot leader made from 3-lb monofilament.

Of course, a good understanding of entomology is also helpful, as is the knowledge of a particular river and the fly patterns that work best on it at different times of the year. However, to get a reasonable handle on what's likely to get you a take, all you really need to do is to try and catch one of the insects once it's hatched – either in flight (difficult, as they're generally quite small at this time of the year), from any rock that protrudes from the river, or by waiting until you see the telltale pulsating circular ripples produced when an adult fly lands on the surface of the water. Scoop the fly out carefully, allow it to dry on the end of your finger and then try to match it with whatever looks closest in your fly box.

Generally speaking, from morning to mid-afternoon, the fly you'll thus be imitating will be a close copy of the emergent adult insect, which in the case of up-winged flies is the first of two forms, known as the 'dun'. Wet fly patterns generally imitate the stage at which the nymph (the larval form of the insect) hatches at the water's surface to become a dun, which then struggles free of the surface film.

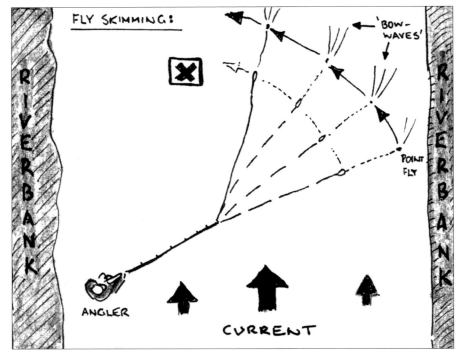

Diagram 2: Fly skimming.

The main idea with wet fly fishing is to present your artificials just below the surface, in the manner of emerging insects, and normally you fish at least two flies, with a point fly and one or more droppers on a leader of at least 6 feet in length. The problem is, that even with two or three flies attached, the unweighted line used to make the leader (sometimes also referred to as a tippet or cast) has a tendency to float, especially when presenting a team of wet flies in the 'across-and-down' mode. This can cause your artificials to skim across the surface of the water (Diagram 2) and will serve as a warning to any wily trout. This problem can be overcome by the application of a mixture of mild liquid detergent and Fuller's Earth to the leader line, a claylike substance that can be bought at any fishing tackle shop, which degreases the line and helps it to sink. The flies can thus be presented in a more natural manner (Diagram 3).

As with a nymph, both the upstream and the across-and-down forms of wet fly fishing are effective for catching trout, although unlike nymph fishing, the downstream version of this discipline relies on more than just the aggressive predatory instinct of the trout. When fished in this manner, the wings on most wet fly patterns will flex as they slide through surface film of the water, imitating the behaviour of a nymph as it begins to shed its larval skin on approach to the surface. The action of the flies slewing across the current as they're held back

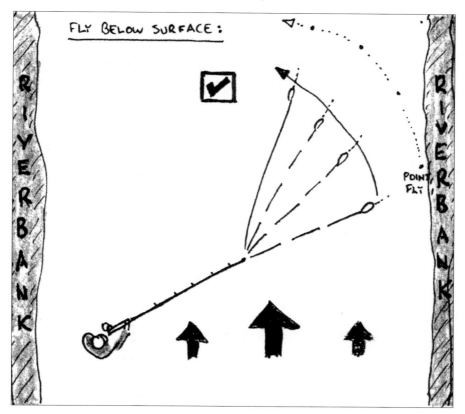

Diagram 3: Fly below surface.

is completely unnatural, of course – but that rarely seems to bother the trout. Nor, for that matter, does the fact that wet flies fished upstream, while moving naturally in the current, slowly sink towards the bottom.

The detection of and dealing with takes in these two forms of wet fly fishing could not be more different, however. When fishing down and across, a take will more often be felt rather than seen, as the trout pulls, usually firmly, against a fairly tight line. Knots need to be secure for this reason and any damaged or wind-knotted leader-line replaced – yet a trout will rarely hook itself and the strike needs to be made instantaneously, as the inevitable amount of curve in the line means the fish is already starting to reject the artificial by the time the take is felt. The longer the line you've cast, the shorter your window of opportunity, so it is worth bearing in mind that it's rarely necessary to cast too far in this sort of river fishing – a fact that should also be borne in mind when fishing stretches of river overhung by bankside foliage.

The cast for across-and-down should be made across the stream, away from the bank you're standing on or nearest to – conventionally at 45 degrees,

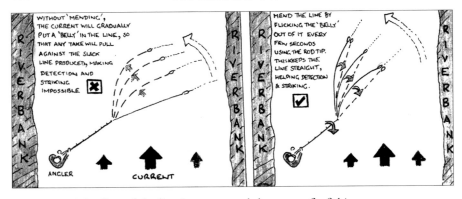

Diagram 4: Mending of the line in across-and-down wet fly fishing.

although I'm quite happy to cast straight across, as this allows me to cover more of the river. The more square the cast, however, and the more 'mending' of the fly line is required, as the curve in it will quickly become unmanageable – that is, you'll eventually be almost unable to feel, and even less able to respond to, any take. Mending (Diagram 4) simply requires the use of the rod tip to flick the 'belly' out of the bend in the line every few seconds (depending on the speed of the current and length of cast), keeping it as straight as possible between the rod tip and flies at all times.

Once the problem of line mending has been overcome, the next dilemma you'll encounter in all forms of river fly fishing is that, unless you've got the cover of a high riverbank or some bankside foliage to keep you off the skyline, you'll often be clearly noticeable to the trout within the 'fish's window' (Diagram 5). This is the view the fish gets of everything within its field of vision above the surface of the river, which although limited to a narrow circle by the refractive index of the water, is enhanced by those same physical properties to give it an exaggerated impression of objects right on the periphery. It follows that a trout lying closer to the surface of the river will see an angler later than one sitting deeper in the water, although the image it sees will be much clearer, and while you'd be surprised how close you can get to a feeding fish – even in the crystal-clear waters of midsummer – the second it does see you, it will be off in a flash. Nonetheless, it should be possible for a stealthy angler, wading carefully as he goes, to get within comfortable casting distance of his quarry, with the fish remaining none the wiser to the fact anything's afoot until the fly has landed on the water. It goes without saying, however, that standing in open water or on an open riverbank with the sun behind you to cast a shadow is something that should be avoided at all times.

Wet fly fishing in the across-and-down style tends to become less effective the longer the season goes on, as the trout become more preoccupied with feeding

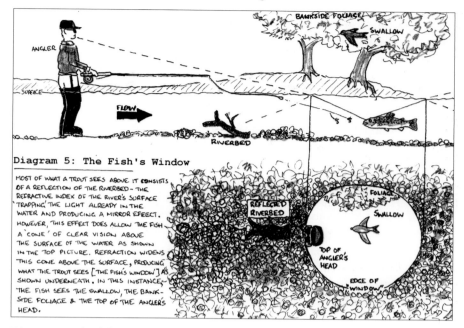

Diagram 5: The Fish's Window

MOST OF WHAT A TROUT SEES ABOVE IT CONSISTS OF A REFLECTION OF THE RIVERBED - THE REFRACTIVE INDEX OF THE RIVER'S SURFACE 'TRAPPING' THE LIGHT ALREADY IN THE WATER AND PRODUCING A MIRROR EFFECT. HOWEVER, THIS EFFECT DOES ALLOW THE FISH A 'CONE' OF CLEAR VISION ABOVE THE SURFACE OF THE WATER AS SHOWN IN THE TOP PICTURE. REFRACTION WIDENS THIS CONE ABOVE THE SURFACE, PRODUCING WHAT THE TROUT SEES [THE FISH'S WINDOW] AS SHOWN UNDERNEATH. IN THIS INSTANCE, THE FISH SEES THE SWALLOW, THE BANK-SIDE FOLIAGE & THE TOP OF THE ANGLER'S HEAD.

Diagram 5: The fish's window.

on flies that have already hatched. However, a wet fly cast upstream will still frequently take trout right through into the summer months. The problem with this technique is that, with the fly travelling towards you below the surface of the water, the detection of takes is entirely different from fishing downstream – and, like the upstream nymph, often imperceptible to the naked eye until you're aware of the telltale signs. The recognition of a take in this form of fly fishing is therefore achieved by watching for movement on the end of the fly line (where it is tied to the leader) and a fish is hooked by striking at any sharp movement against or across the current. It is also necessary to stay in touch with the fly by retrieving line with the free hand in order to be ready to respond instantly to takes. In a run with a number of trout in residence, both upstream and obliquely across from you, these can happen at any time – although casting to a seen fish that accepts the offer should result in a take within seconds.

The upstream wet fly technique is thus like a cross between upstream nymph and dry fly fishing, in so far as the takes are seen through the movement of the line rather than the physical action of the fish, yet the fly should normally only be on the water for a second or two before it gets taken. This can make for heart-stopping sport, but it also means that you'll probably only get one chance at a seen fish before you put it down!

There are, of course, any number of nymph and wet fly patterns that are a decent bet on most trout streams in the North East, given their ability to closely imitate a good number of different insect species. The Greenwell's Glory is one such example, as are the Pheasant Tail and Gold Ribbed Hare's Ear – either of which will work well in both their un-weighted nymph and wet fly forms – while the Partridge & Orange and the Snipe & Purple are each good all-round flies for North East trout fishing from April onwards. All of these patterns are intended to imitate a variety of olive species, but there are two species in particular that, given their prevalence on many northern rivers during spring, are out and out killers in their exact form. The Large Dark Olive and the Iron Blue Dun can both be prolific at this time of the year, with the former abundant on, among others, the River Derwent, while the latter is more commonly encountered on Northumberland streams such as the Wansbeck.

At the end of May and on into early June, on most rivers in this area, trout anglers will witness the most spectacular entomological event of the year in form of the mayfly hatch. Mayflies are large up-winged flies of the same biological order as the olives (*Ephemeroptera*), which can hatch in their hundreds, and sometimes thousands, from even the smallest bodies of water. The sizes of these hatches vary naturally from river to river, and year to year – as do the types of mayfly, although by far the most common species in the North East is the characteristically grey and white *E. danica* – but there is one constant in all of this: brown trout go absolutely crazy from them! This can be both a blessing and a curse, as while it eliminates all but the most rudimentary of fly choices, trout can become so preoccupied with the mayfly that they won't so much as look at anything but an exact imitation of the natural.

Like all up-winged flies, mayflies have three distinct life stages – the first two being the nymph, which lives on the riverbed for about a year before rising to the surface and hatching into the dun. Unlike other aquatic fly types, however, this emergent up-winged fly is not sexually mature and it therefore undergoes a further change (which in some cases includes a complete alteration in body colouration) into the imago, or to use the more commonly spoken angling term the 'spinner'.

As mayflies are extremely short-lived in their adult form, the spinners court and mate close to and above the river they hatch from, before the females return to the surface of the water to lay their eggs and die. The response of the trout to this is entirely predictable, and although they will feed heartily on those insects hatching in the surface film, in reality they have little need to, due to the banquet of spent flies that subsequently cascades back down onto the river. Almost as awe-inspiring as watching vast columns of duns forming a haze above the water after hatching is seeing the spinners falling back onto the surface. The rhythmic circular ripples produced by the flies in their death throes are easy to see – and even more noticeable to the trout beneath the surface. It's rarely more than a few seconds before a much broader set of rings starts radiating out from where the mayfly once was!

The blessing is that in their frenzy to consume as many adult flies as possible, the trout's normal reticence often goes out of the window once the mayfly hatch gets into full swing. Provided you've got a pattern that resembles fairly closely the type of mayfly that's hatching, you're unlikely to encounter any of the fastidiousness shown at many other times of the year. As previously mentioned, by far the most common mayfly found on Northumbrian rivers is the white and grey coloured 'danica', and while classic patterns such as the Grey Wulff will make good imitations, a look through the fly cabinet at any decent tackle shop is likely to produce something similar that will work.

Nonetheless, presentation is still paramount, even during the mayfly season. It goes without saying that a dry fly is the only thing that will properly imitate a spent mayfly floating on the surface of a river, and while a wet fly approach may interest some trout at the precise moment of a hatch, they are unlikely to preoccupy themselves with emerging insects once the veritable banquet of easy pickings starts dropping down onto the water. Dry fly fishing presents its own unique challenge to the trout angler, combining one or two of the skills already discussed, but although the basic principles of upstream nymph and upstream wet fly fishing do hold true for dry fly fishing, there are two very important differences.

Firstly, the fly (you only use the one in dry fly fishing) has to be presented floating on the surface of the water and must not sink, while the leader – which is clearly visible to the fish if *it* floats – must sink into the surface film. Flotation of the fly can be achieved by use of an aerosol spray commonly termed 'gink', several makes of which are available at most fishing tackle shops, while the leader can (again) be treated with Fuller's Earth to help it sink. Equally importantly, in order to present a dry fly properly, it must normally be cast either directly or (more usually) at an angle upstream for reasons that will become obvious – a dry fly fished downstream will not work for anything other than a very accurate cast to a fish that takes instantly. And a floating fly cast upstream on the end of a line gives rise to another issue, as the fly must travel naturally *with* the current, or else the trout will all too easily spot the deception. The fly must not skim along the surface at a pace quicker than that of the current or be slowed down so that it drags, something that may seem fairly straightforward but is, in practice, quite difficult to overcome.

This problem is quite appropriately referred to as drag (Diagram 6) – the drag that the fly line and (in particular) the leader impart on the floating fly, owing to the different speeds at which the current will be moving on different parts of the surface of the river. Drag can rarely be avoided – except when you're casting to a position directly upstream, a practice avoided by many dry fly anglers for fear of 'lining' (spooking the trout when the fly line alights on the water above its head). Even then the natural quirks of most trout streams will impart *some* drag.

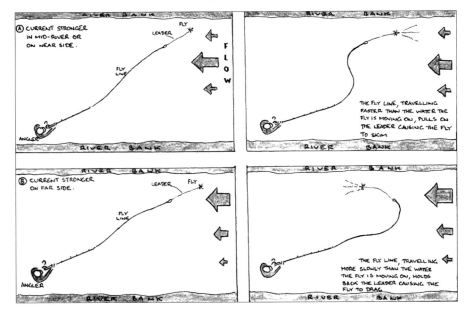

Diagram 6: Common types of drag in dry fly fishing.

Drag has therefore to be compensated for and this is achieved in a manner similar to mending the line in across-and-down wet fly fishing. Indeed this is necessary, to a less critical degree, in upstream wet fly and nymph fishing too (in order to keep the fly line as straight as possible) but in dry fly fishing the 'mend' is far more tricky, as it has to be made in such a way as not to impart any further drag upon the natural movement of the floating fly.

The more experienced angler can also compensate for drag by making a 'wiggly cast' (Diagram 7) – applying extra force to the forward cast that delivers the fly, in such a way that the line shoots forward to become fully extended before it hits the water, springing it back and creating kinks as it alights on the surface. This obviously requires prior knowledge of the differential flow-rates on any given pool, as well as no small amount of practice at getting the forward cast just right so as to produce a 'wiggly line' that will straighten as it travels downstream.

The only way to master these techniques is by practice and probably the best way is to begin by fishing in fairly slow-moving pools where the overall current on the surface is more even. Once you've mastered these, you should then be ready to move up to more challenging, quicker (or differentially) flowing pools, and with this in mind it's worth considering another odd quirk of dry fly fishing. The temptation for many beginners might be to stick to fishing the 'starter' pools in these circumstances – especially if one or two trout have been taken – but there are good reasons why those more challenging places are

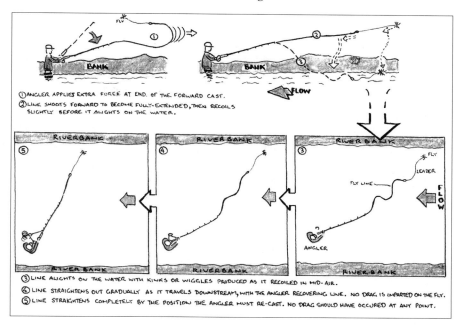

① ANGLER APPLIES EXTRA FORCE AT END OF THE FORWARD CAST.
② LINE SHOOTS FORWARD TO BECOME FULLY-EXTENDED, THEN RECOILS SLIGHTLY BEFORE IT ALIGHTS ON THE WATER.

③ LINE ALIGHTS ON THE WATER WITH KINKS OR WIGGLES PRODUCED AS IT RECOILED IN MID-AIR.
④ LINE STRAIGHTENS OUT GRADUALLY AS IT TRAVELS DOWNSTREAM, WITH THE ANGLER RECOVERING LINE. NO DRAG IS IMPARTED ON THE FLY.
⑤ LINE STRAIGHTENS COMPLETELY BY THE POSITION THE ANGLER MUST RE-CAST. NO DRAG SHOULD HAVE OCCURED AT ANY POINT.

Diagram 7: The avoidance of drag using the 'wiggly cast'.

worth a go. First is the fact that such spots are often favoured by bigger trout and they can also often be the only places you'll get a take in conditions of low water, owing to the fact that trout in the slacker areas have more time to inspect (and frequently reject) the offer of an artificial dry fly.

Another infuriating complexity of dry fly fishing for the beginner is the tendency (when using a leader of an even thickness, or gauge) for the fly to be consistently delivered to the wrong position on the water in anything greater than a moderate breeze. Typically, in all but the most extreme conditions, the end of the fly line will travel to exactly the place you intended it to, only for the fly to be blown to one side, landing a yard or two to the right or left of where you actually want it. The problems of this are twofold – firstly, of course, the fly is wrongly placed for it to be seen by the trout you are casting to and, secondly, the leader will be at an angle to the fly line, presenting major problems in both the control of any drag and the ability to respond to any take that does occur.

This problem is caused by the fact that, as the fly line is thicker and much heavier than the leader and the fly on the end of it (dry flies are by nature lighter than wet flies and nymphs), the force that turns the fly line over as it travels through the air dissipates as soon as its whole length has alighted on the water, leaving no power to thrust the leader and fly onwards. The fly is thus all too easily deflected by the breeze, resulting in the situation described above.

The answer to this is the use of a tapered leader, which can be either bought ready-made as a length of nylon monofilament that is of a thicker gauge at the end you tie it to the fly line and narrower at the end you attach the fly, or 'constructed' from one or more pieces of level mono. The tapered leader, being of a far closer thickness to the fly line itself at the point they are joined, better facilitates the transfer of the force that turns the fly line over – into and on down the leader, so that the fly is more likely to be delivered to the right place.

The construction of your own tapered leader can be achieved by tying together several lengths of different gauge monofilament, using a series of four-turn water knots. Going from the 'thick' end down to the point, a 9-foot leader (for example) could consist of 36 inches of 10-lb BS, then 24 inches each of 8-lb, 5-lb and finally 3-lb BS mono. The whole thing, when attached to the fly line at the blunt end should behave exactly like a tapered leader; however, care should be taken to avoid attaching directly together any two lines with breaking strains of greater than a 2:1 ratio, as the modulus of the stronger line (the force applied as it contracts after stretching – a property characteristic of nylon mono) is likely to break the weaker line on the strike. In the above example, the 8-lb and the 5-lb sections would 'cushion' the 3-lb tippet from the 10-lb top section.

This brings us to the last technical feature of river dry fly fishing – the take – and more pointedly the rate at which the angler should strike. The take in dry fly fishing is, of course, unmistakable as the fish breaks surface in a characteristic 'rise' and takes the artificial. The rise is, needless to say, the definitive indication of a take and although movement of the fly line will often be seen, as we will see, this should not be regarded as the thing to watch out for. Rather, it is necessary to watch the floating fly as it travels downstream and while there is the obvious temptation here to strike at the moment a trout rises to take the artificial, this should be avoided as it will as often as not pull the fly clear of the fish's mouth before it has been able to properly take hold of it. Conversely, if you wait too long – and this would often mean that you saw the end of a 'wiggly' fly line move – in all probability the trout will have realised its error and rejected the fly. Therefore, the timing of the strike is critical and it mustn't be either too fast or too slow for the above reasons. Stillwater anglers say that you should allow sufficient time to utter four syllables, with the easily remembered examples being either 'God Save the Queen' or 'Good Morning Trout', before setting the hook, although from my own experience, two syllables – i.e. 'God Save ...' or 'Good Mor–' – or even quicker, is about right on rivers. As with all fly fishing techniques, however, practise is the key!

Of course, river fly fishing for trout doesn't just end following the two weeks or less of the mayfly season – you can carry right on throughout the summer and up until the season ends on 30 September. It goes without saying that different species and different types of flies come into play at different times of the season and, with regard to olives, mid to late summer days can see an abundance of the blue winged olive (in the afternoon and early evening) and its imago the sherry spinner (evening) on many northern streams.

Sometimes, while up-winged flies will still hatch in lesser numbers, aquatic flies from other well-known biological groups will predominate – often causing the trout to switch to feeding exclusively on these insect forms. Common culprits in these instances are members of the order *Tricoptera*, commonly known as caddis flies or 'sedges', which are characterised by their roof-winged profile, when at rest, as well as the presence in many cases of long antennae on their heads. In flight they are usually brownish in colour when seen to the naked eye and, in contrast to the ascendant flight of the olives, they tend to whizs round randomly in swarms just above the river surface. A caddis or sedge imitation correctly matched to the adult fly will stand the best chance of temping a trout, although it will sometimes be fooled by an up-winged artificial of the same size and colour cast onto faster water.

Another type of fly that can frequently be seen flying randomly in swarms just above the surface of many northern streams in spring and summer is the much smaller needle fly, which belongs to the order *Plecoptera*, or the 'stone flies'. Needle flies are so called because their wings furl around their greyish-coloured bodies at rest – catch one and let it sit on your finger and you'll see – and although there are artificials that represent exact copies of these insects, I've usually enjoyed success when these flies are on the water by fishing with a small grey duster.

Lastly, although usually considered of greater importance to the stillwater fly fisher, midges are an obvious group to which particular attention should be paid, although imitations of this order, the *Chironimidae* are generally fished either in the surface film or below it, rather than as true dry flies.

Although river fly fishing for trout generally involves trying to catch just the one species of fish, the challenges it presents to the angler are many. Mastering it can take a lifetime, so it's little wonder that so many North East fishermen remain faithful to its charms. If you've never tried fly fishing before, why not give it a go this year – you might just find yourself hooked for life too!

CHAPTER TWELVE

Summer: Coarse Fishing Techniques for Stillwaters and Rivers

16 June is an iconic date for most traditional coarse anglers, although nowadays very few in the North East even realise its significance. For several decades now, a dispensation rule relating to the close season on stillwater venues means that the majority of North Eastern coarse anglers – with their preference for this form of the sport – have never needed to know the time-honoured opening day of the season.

Nevertheless, the middle (or at least the start) of June still marks the beginning of the best time of year for most forms of coarse angling and while, to be truthful, fishing for species such as chub and pike may well be better in winter, both can still be relied upon to provide outstanding sport from June to October. The same cannot be said for the tench, carp or barbel, however, and given the tendency for these species to disappear off the angler's radar at the first sign of a nip in the air, the warmer months usually comprise his or her only real chance with these species.

Of course, in purely aesthetic terms, coarse fishing in the summer is simply a more comfortable and enjoyable pastime, but another advantage it has at this time of year is that its two forms – river and stillwater – are generally at their best under differing weather conditions. In game fishing, a sunny midsummer day can often mark the onset of 'dog days' on running water, while bright weather is well known to put rainbow trout down on lakes and reservoirs. Rain, meanwhile – with its attendant flooding effect – spoils river fly fishing completely, with some streams often taking days to recover. And a rainy day is nothing better than a nuisance to the stillwater game angler.

River coarse fishing, by contrast, tends to be at its best a day to 48 hours after heavy rain, as the colour drops out of the ebbing floodwater and the river begins to fine down. In a normal North East summer, therefore, sharp spells of rain followed by a day or two of dry weather are optimum for anglers looking to catch species like chub and barbel. It follows therefore that low water, clear skies and blazing sun are conditions the river angler should try to avoid, yet on stillwaters, these can be the sort of days you dream about! On these

A classic stillwater swim at Brasside Pond in County Durham. The author has caught numerous tench and the odd carp here by casting a slider float rig as close a possible to the lilies.

venues, steady rain is the thing you should try to steer clear of, coinciding as it does with cooler air and water temperatures, and while such conditions don't necessarily make fishing the lakes and ponds a futile endeavour, a sunny or muggy day will undoubtedly be better. This means that while one type of venue is 'off', the other might very well be 'on', so a versatile approach can often be the key to a successful summer's fishing.

Beginning on still water, then, we can assume that the weather is fair and warm, and that any rainfall that is forecast won't amount to more than the odd rogue summer shower. The breeze at this time of the year in lowland areas of the North East will rarely amount to more than the odd gentle gust, but unless you've got a liking (or a hot tip) for a particular hotspot, the first rule is to head for that part of the lake or pond that's facing *into* any breeze. The reason for this is that the windward bank is the place where most of the food the fish will be foraging for will either have been blown or drifted to on the prevailing wind-driven current. A feature – any kind of overhanging tree, branch or piece of bank, a headland or bay, or even a reed-bed – is the next thing to look for, and while not all stillwaters have these, an area of deeper water may prove advantageous. If you can't find deeper water, however, don't despair, as places like park lakes (which rarely have such features) still often provide many a specimen carp to the specialist angler, as well as large bags of silver fish for the pleasure fisherman.

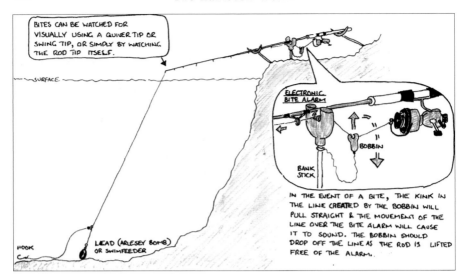

Diagram 8: Stillwater legering – general principles.

Moving on to tactics, there are two distinct ways in which to fish a stillwater, leger or float, and while the former (see Diagram 8) is more straightforward, float fishing is by far the more versatile, as it can cover almost all the situations legering does and more besides.

Floater fishing (see Diagram 9) is the only form of specialist carp fishing we will cover here and is a technique which profits from the carp's liking for floating baits fished on the surface on a warm, still day. The concept is simple: cast out a piece of bread crust or a water-softened dog biscuit on either a free line or with a controller float to aid distance and wait for a hungry carp to come along. In practice, it's never that straightforward, however, as carp will never, or hardly ever, visit certain parts of a pond and, in those places where they do, they'll often patrol around on a highly repetitive course. The trick is to sit for a while and watch for carp feeding in this way and then to try and observe their patrol 'route' without them being able to see you. Having got this far, catapult a few free offerings into the most appropriate area into which to cast – just *before* the patrol (which usually consists of a number of fish) arrives. If the offerings are accepted, repeat the process on the next pass, this time also casting your baited hook into the middle of the floating armada of bait. Takes will be unmistakable, but always be certain to remember which floater is the hook bait! A controller will give some indication in this regard, but it will also impart a degree of drag to the carp's take – giving it an earlier warning of its mistake than would be the case with a free line. Tackle for this type of fishing should typically consist of a Mk VI carp or a strong avon type rod of at least 1¼-lb test curve, 8–10-lb breaking strain line and size 6 hook or larger.

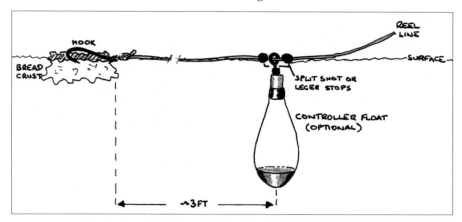

Diagram 9: Floater fishing.

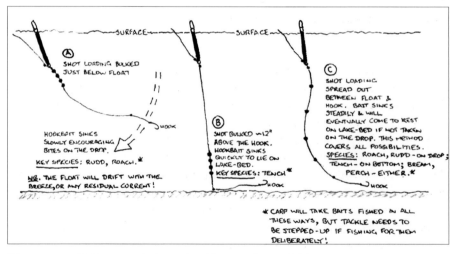

Diagram 10: Different techniques in stillwater floater fishing.

Fishing 'on the drop' is more appropriate to smaller species such as roach and rudd, which tend to inhabit the surface layer of a pond or lake in summer, however this technique will also take carp on stouter tackle. The principle here is to place all the weights necessary to cock the float immediately below it, allowing the hook-bait to drop slowly through the surface film (see Diagram 10). On a day with even the slightest breeze, the float will tend to drift, allowing the angler to search out a larger swathe of the upper layers of the water, and taking a measure of the depth is unnecessary as the idea is only for the bait to sink a few feet at most. Any loose-feed or ground-bait should thus be aimed *ahead* of the float's trajectory (Diagram 11), so it will already

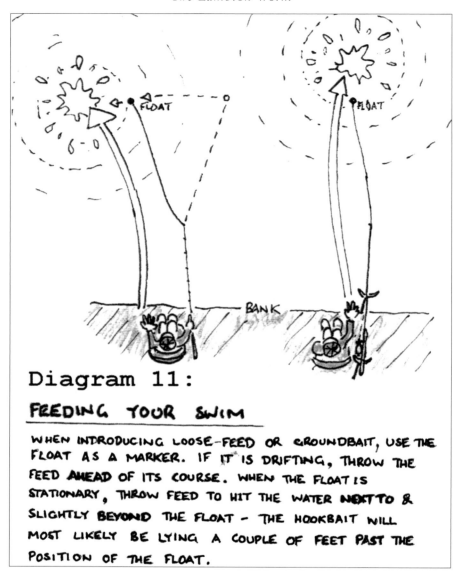

Diagram 11:

FEEDING YOUR SWIM

WHEN INTRODUCING LOOSE-FEED OR GROUNDBAIT, USE THE
FLOAT AS A MARKER. IF IT IS DRIFTING, THROW THE
FEED AHEAD OF ITS COURSE. WHEN THE FLOAT IS
STATIONARY, THROW FEED TO HIT THE WATER NEXT TO &
SLIGHTLY BEYOND THE FLOAT — THE HOOKBAIT WILL
MOST LIKELY BE LYING A COUPLE OF FEET PAST THE
POSITION OF THE FLOAT.

Diagram 11: Feeding your swim.

have reached the depth the fish are feeding by the time the hook-bait arrives.
Appropriate tackle for this method, when fishing for roach and rudd, would
be a match rod, 2–3-lb BS reel line, with a hook of size 16 or smaller tied to a
1–2-lb hook length. Baits for these smaller species can include maggots, small
worms, bread and corn.

We now come to those stillwater float fishing techniques that correspond
to leger fishing (i.e. the bait is fished on the bottom of the lake or pond) and

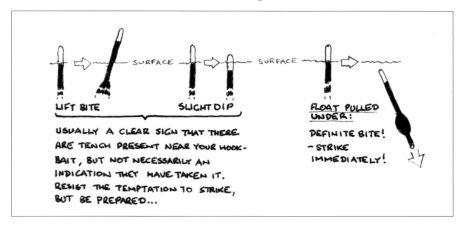

Diagram 12: Typical tench bites.

although they are much trickier to master and require considerably more setting up, they are advantageous to legering in three important respects. Firstly, they eliminate the presence of a taught (usually diagonal) line that exists in your swim when you leger or feeder fish, a situation that can lead to fish spooking if they repeatedly bump into it (i.e. line bites). Back leading is a solution used by carp anglers fishing at long range, but at close quarters, float fishing is by far the better option. The line from the float to the lakebed will be almost vertical and presents no more resistance to the fish than what they would feel when swimming into a reed or lily stem, and when a fish brushes the line with its body or tail it will produce a very distinctive line bite (see Diagram 12) that won't cause it to spook.

Secondly, the presence of a float gives a very exact indication as to the whereabouts of your hook-bait, allowing you to very accurately introduce loose-feed or ground-bait without spreading it out all over the lakebed in the general vicinity. Also, given that bottom-feeding tench and carp produce telltale gas bubbles that effervesce on the surface of the lake or pond immediately above them, the presence of your float with respect to this phenomenon can tell you whether to be ready for a positive response imminently or in a short while.

Lastly, and some would say most importantly, in classic stillwater legering the rod is usually pointed downwards and if you are using a quiver or swing tip as an indicator, a bite is very easily missed unless you are looking down at it all the time. With a float, you are looking out across the water with far greater freedom to glance around at the scenery or go about the business of tackling up a second rod or suchlike. Furthermore, a bite indicated by the dip of a float is far more pleasing to most anglers than the sound of an electronic alarm.

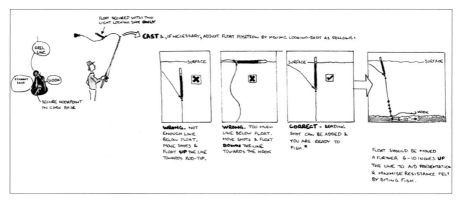

Diagram 13: Setting the depth of the float.

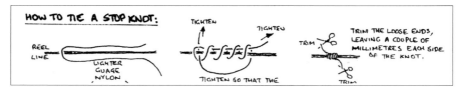

Diagram 14: How to tie a stop knot.

Of course, with float fishing on the bottom, taking an accurate depth setting in advance is essential (see Diagram 13) and you will need a plummet lead for this purpose. In the North East, there are very few coarse fishing venues that have the sorts of depth found further south in gravel pits and the like, the kind that would banish any thought of float fishing on the bottom. About the deepest I know is Dissington Pond, a former small limestone quarry near Ponteland that ranges from 10–15 feet. These sorts of depth will, however, still cause float anglers trouble with their casting and may make some think seriously about reverting to leger or feeder fishing. This is where the slider float technique comes in – a method specifically designed for coping with depths in the range of 10–20 feet.

Instead of employing a float fixed onto the line with spilt shot or leger beads, this technique allows the float (any normal float will do, except that it mustn't be self-cocking) to slide up and down the line and be controlled by a stop knot that will cast through the rod rings and wind on and off the reel when the float is out of the water. Tying the stop knot is easy (see Diagram 14) and the rig it produces allows the length of the line you need to cast from the bank to be as little as two feet, while the float itself behaves in exactly the same way as a conventional one with the exception of the first few seconds after the cast (see Diagram 15). Taking a depth setting for the slider is essential, but can

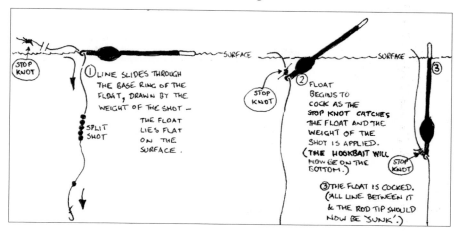

Diagram 15: How a slider float will behave after casting.

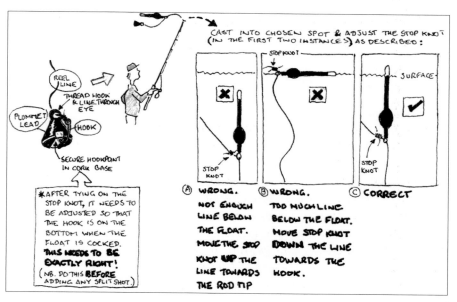

Diagram 16: Setting the depth of the float in slider float fishing.

be achieved in pretty much exactly the same way as with a normal float (see Diagram 16).

Tackle for tench fishing using either a conventional or slider float rig can vary. Where carp are present, it should be stepped up to 6-lb BS line, minimum, in tandem with an avon or carp rod, as described previously, whereas on waters where only tench are found a 4-lb line should be sufficient, fished with a match or float rod. Hook lengths can be included, but fishing with mainline 'straight through' will make setting up a slider rig less problematic. Common baits for tench include corn and luncheon meat, but these can become 'blown' on heavily fished waters – at Dissington Pond in the 1990s, anglers counteracted this by using fresh cockle. Boilies and pellets will also work and hair rigs can be included as shown in Diagram 22.

* * *

So we come to the running water part of our North East summer coarse angling extravaganza, and while this particular discipline may indeed be the sole official preserve of that old tradition associated with 16 June, in actual fact, from mid to late June, stillwater species such as the tench are normally a far safer bet. The problem in these more northerly climes is that late June (and sometimes even early July following a cold spring) marks the spawning season for many river-dwelling coarse species, and therefore, while there may well be catches made during the early weeks of the season, in most years the more consistent summer fishing doesn't usually commence until the first or second week of July.

Up here in the far North East of England there is a further distinct split in the quality – and indeed the availability – of river coarse fishing. With the exception of the Tweed, which contains roach and dace in its lower reaches, none of the rivers north of the Tyne catchment contain any appreciable number of coarse fish. The main River Tyne itself does have roach, dace, chub and perch (as well as the occasional 'rogue' pike); however, its principal North and South tributaries, and the feeder streams thereof, are largely devoid of them. The Tees and Wear contain a similar range of non-salmonid species to those of the Tyne, in addition to the bream and the barbel, a hard-fighting native of the Yorkshire Ouse river system introduced surreptitiously by local anglers during the 1970s. Further south, the Swale and the Ure, in common with all tributaries of the River Ouse, contain all these fish quite naturally, and the quality of the coarse angling here is generally of a higher standard than on their larger Northumbrian neighbours.

As has already been mentioned, the ideal conditions for river coarse fishing are those that follow regular spates caused by reliable summer rainfall. Even unseasonable heavy rainfall, such as that experienced during late June and early July 2007, can be the river angler's friend. That was a season in which I recorded some of my best chub fishing ever on the River Wear, with only the highest flood conditions unyielding and the sport frenetic from the end of June right through to mid-September. Drought conditions are not good for river fishing, however, and the angler might be best sticking to the ponds and lakes in these circumstances.

Like in stillwater angling, there are two distinct forms of coarse fishing on rivers – float, which allows you to probe a swim by running a bait though

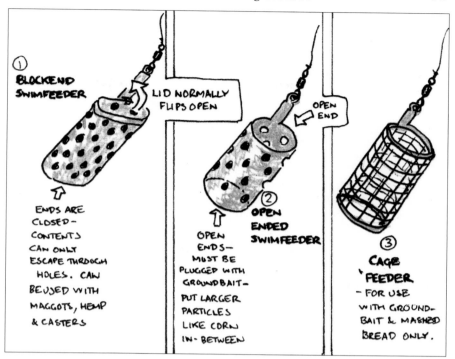

Diagram 17: Types of swimfeeder.

at variable depth on the current, and legering, which places a bait hard on the bottom of the river that can be either static or be 'trundled' downstream gradually. This latter discipline has been further refined to produce probably the most popular form of modern river fishing, through the substitution of the leger weight or 'bomb' by a swimfeeder that trickles a steady flow of particle loose-feed or ground-bait around the hook-bait. This now comprises the starting point for most anglers who coarse fish on northern rivers, with standard practice being the setting up of two rods – one for legering or feeder fishing and a second for float.

We'll begin with swimfeeder fishing and despite the fact that this clever little device allows you to introduce ground-bait or loose-feed to the fish in your swim on every cast (feeders come in block-end, open-ended and cage varieties – see Diagram 17), it is worth bearing in mind that to get them there in the first place, it's generally a good idea to pre-bait your chosen swim. Alternatively, if you're lucky and you've got a whole stretch of river to yourself, you could pre-bait several swims. For barbel and chub fishing, my favoured pre-baiting routine involves the introduction of two or three pints of cooked hempseed (boil the raw seed in water the night before for three quarters of an hour until it splits – any spare or left over can be frozen), along with about half a pint

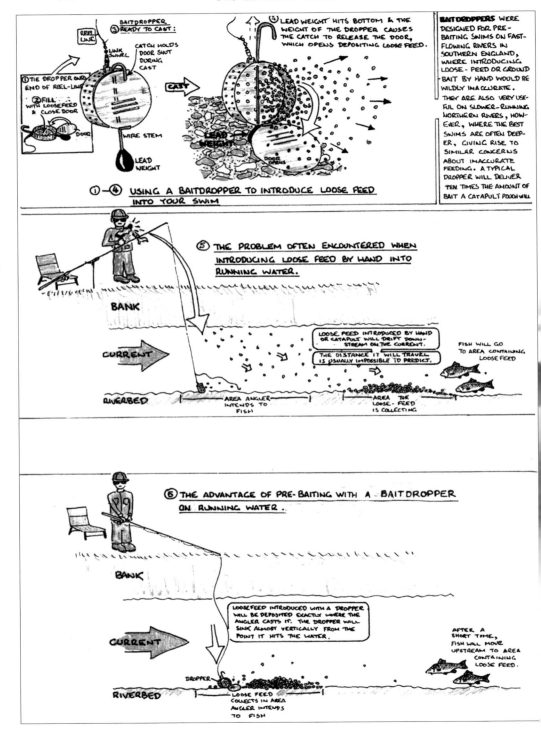

①—④ USING A BAITDROPPER TO INTRODUCE LOOSE FEED INTO YOUR SWIM

BAITDROPPERS WERE DESIGNED FOR PRE-BAITING SWIMS ON FAST-FLOWING RIVERS IN SOUTHERN ENGLAND, WHERE INTRODUCING LOOSE-FEED OR GROUND-BAIT BY HAND WOULD BE WILDLY INACCURATE. THEY ARE ALSO VERY USE-FUL ON SLOWER-RUNNING NORTHERN RIVERS, HOW-EVER, WHERE THE BEST SWIMS ARE OFTEN DEEP-ER, GIVING RISE TO SIMILAR CONCERNS ABOUT INACCURATE FEEDING. A TYPICAL DROPPER WILL DELIVER TEN TIMES THE AMOUNT OF BAIT A CATAPULT POUCH WILL

⑤ THE PROBLEM OFTEN ENCOUNTERED WHEN INTRODUCING LOOSE FEED BY HAND INTO RUNNING WATER.

⑥ THE ADVANTAGE OF PRE-BAITING WITH A BAITDROPPER ON RUNNING WATER.

Diagram 18: How using a bait-dropper can help on running water.

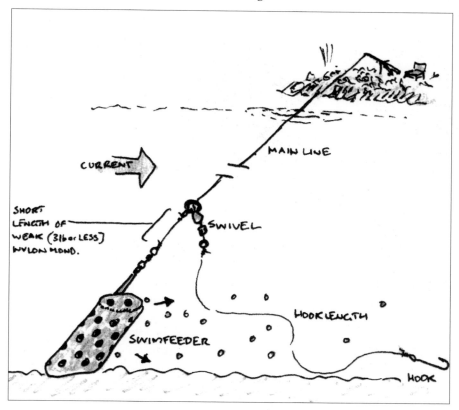

Diagram 19: River swimfeeder fishing – general principles.

of either maggots or casters. Corn, if you intend to fish with it, will also need to be introduced (sparingly – no more than a sprinkle or you'll overfeed the swim), though luncheon meat or bacon grill need not necessarily be pre-fed, as hempseed tends to be a trigger for fish to feed on this bait in itself. In modern angling, it is also impossible to ignore fish or 'halibut' pellets, protein-packed pill-like baits that can be bought at most tackle shops. A standard bag of small pellets will provide several 'feedings', with my method to add about a pint on top of the rest of the feed, whether or not I intend to use larger pellets as a hook-bait. The whole lot is introduced by use of a bait-dropper (see Diagram 18), a device that offers the angler a considerable advantage over pre-baiting by catapult or by hand insofar as it places all the feed exactly where you intend to cast your hook-bait.

Once your swim has been pre-baited, the textbook approach is to wait for half an hour or so to allow it to 'settle', or in other words to allow the fish to regain their confidence after the commotion of pre-baiting. In reality however, from my own experience the time required to go about setting up your swimfeeder rig (Diagram 19) is normally sufficient for any barbel or chub that were in the vicinity to have returned and begun feeding confidently, although a longer period may well be needed when fishing for smaller species like dace and roach. This will obviously vary in different places and on different days, but there is no harm in casting a line, as if the fish have returned promptly, you might as well be fishing for them as sitting and waiting.

The tackle required for feeder fishing for the larger species in summer is governed by the likely presence of barbel on most of the North East's coarse fishing rivers, namely the Wear, Tees, Swale and Ure. While the summer chub – and big river-dwelling bream, where they are found – can quite safely be fished for on comparatively light gear, generally speaking, even modest-sized barbel will put up a fight out of all proportion to their size, and a big fish making use of a strong current can easily out-muscle a carp, weight for weight. Given that barbel, chub and bream are particular to most of the same baits, and that in the first two instances they are also frequently found in the same kinds of swim, an avon style rod of 1¼- or 1½-lb test curve is therefore essential, combined with 8-lb breaking strain mainline. Hooklengths can be made up from either nylon or braid (see Diagram 20) and should be of a minimum 6-lb BS. While this sort of a set up may be considered over the top for chub fishing, in addition to the barbel, it will give extra confidence in the likely event of your hooking one of the many chub weighing 5 lb or more that are found in these rivers, fish which can put a fair amount of pressure on inadequate tackle.

Feeder fishing for the smaller coarse species with maggot, worm or corn (as well as for trout and grayling) can, however, be undertaken using lighter tackle. A normal leger rod combined with a 3- or 4-lb mainline will suffice in these situations, although the temptation to use any hooklength with less than half the b.s. of the reel line should be avoided.

Having sorted out tackle and baited up, the question of where exactly to cast your swimfeeder (and indeed where to concentrate your pre-baiting) is a moot point for many anglers. Obvious summer hot spots for summer chub, and in particular barbel, include any place with a fairly strong current and deeper water, for example bends (which usually provide both these features on the outer bank) and weir pools (likewise, below the sill). Faster, shallower water is not typically a good place to look for those larger species, but can often be a productive area for dace, when the depth of the water makes stick float fishing problematic. Roach, and especially bream, will usually favour slower-flowing water, which in its very nature is normally fairly deep.

Of course there are exceptions to this. Any angler who has ever fished Feren's Park on the Wear at Durham City will immediately see the shortcomings of these river fishing stereotypes, for while those hard and fast rules are a good general guide to where to fish, there is crossover and sometimes there are places that flagrantly break them. Feren's Park is a slow-moving stretch of

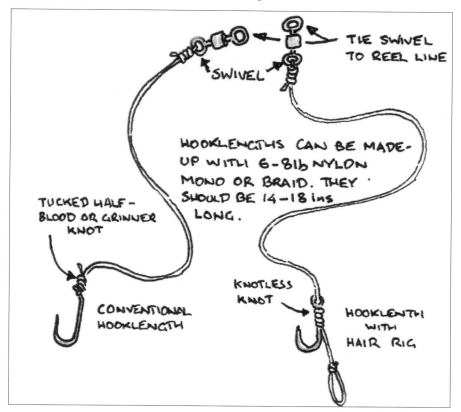

Diagram 20: General hook length for leger & feeder fishing.

river with a 'stepped' profile introduced by the regulatory authorities with the aim of flood management (similar features are also found throughout the navigable reaches of the rivers Ouse and lower Ure in North Yorkshire). The Durham version features a 300-yard-long mid-river channel, six to eight feet in depth, fed at its top end by a swift shallow glide and gathering pace again as it shallows up on approach to a sharp bend.

On the face of it this looks like classic bream fishing water – and it is, with many a large slab having been caught from its prominent right bank pegs down the years; however, this place is even more hallowed as a barbel hotspot, as well as being highly productive on its day for specimen chub, perch, dace and the occasional monster pike. Unfortunately, like all well-known venues it is prone to being overfished and can therefore appear 'unproductive' once all the common baits and techniques have been blown. The solution can sometimes be to fish at night, although in common with most places the ability to choose a dull, overcast day over a bright one will bring about an instant upturn in your chances.

Diagram 21: Rolling or 'trundling' a legered bait.

Such is the almost universal use of the swimfeeder in river fishing these days that, other than tight swims on smaller streams (a description that none of the main North East coarse fishing rivers fit), straightforward legering tends only to be practised when an angler wants to 'trundle' his bottom-fished bait through the swim. Owing to their shape and bulk, swimfeeders clearly preclude this technique and leads intended for use in this way should always be round in shape to enable them to roll – or more likely, on typical rock-strewn northern riverbeds – to be 'bounced' along the bottom in the current. The best types of 'ordinance' to employ in this instance are drilled bullet leads or arlesey bombs, with the key to use just enough weight to hold bottom, such that either the building pressure of the current on the line or the lightest pull on the rod tip will dislodge the lead and make the bait trundle off again to find its next resting place, as though being washed naturally downstream.

The main advantage of this method is in its ability to probe a swim on each cast, trundling the hook-bait over a pre-baited area of the riverbed, and taking the bait to the fish, rather than waiting for them to come to it. The more natural aspect of this technique's presentation is also obvious, especially in a swim with a stronger flow, and the fact that you hold the rod for most if not all of the time gives rise to its most commonly used description, 'touch legering', as you feel for bites with the rod held in your hand (Diagram 21) – sometimes also holding the line between the thumb and forefinger of your free hand for still greater sensitivity.

Another slight variation on this method is the feted 'upstream leger' technique, in which the bait is cast upstream and bites are identified by any slackening off on the line – the 'drop back bite' – very difficult to detect. The

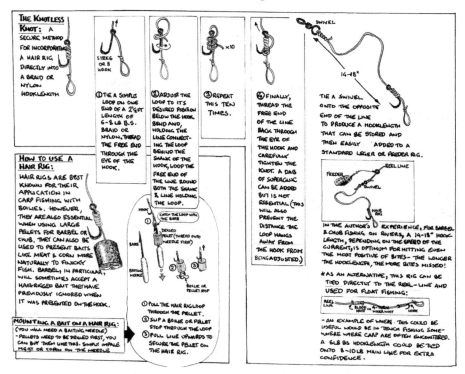

Diagram 22: How to make and use hair rigs.

most obvious places to employ touch legering are those with a stronger flow, such as at the head of a deepening pool where a section of fast-moving water breaks into slower water or forms an eddy, while, by its nature, upstream leger works best in less vigorous currents.

* * *

There are any number of baits that work well with the leger and swimfeeder methods described, some working better than others, while some are good reserve baits when all else has failed. As stated previously, it is impossible to ignore fish (halibut) pellets in modern angling, baits which, in themselves, have built-in fish-attracting properties. The different sizes in which they come obviously work well with different-sized species – smaller ones, which can be attached to a conventionally tied hook with a bait band (these can be bought with the pellets) often accounting for roach and smaller bream. Larger ones need to be drilled (although they can now be bought pre-drilled) and attached to the hook by means of a hair rig (see Diagram 22). This is now the favoured bait of most specimen barbel anglers and it will also work just as well with chub – if you use a swimfeeder, fish with small pellets in a 'block-end'. Use hook sizes 14 or less with a bait band, sizes 6 or 8 for hair rigs.

Moving down the list of the most popular baits, luncheon meat (bought from the supermarket, although flavoured varieties can be obtained at tackle shops – bacon grill, spam and chopped pork also come under this description) is a highly selective bait for barbel and chub and has been for several decades. It still works as well as ever on all but the most heavily overfished venues (places where pellets will have been seen just as often by the local residents!) and is my first reserve on most chub and barbel outings. It is most commonly fished on the hook using a grass stem inserted through the bend to stop it falling off, but can be fished on a hair rig in the same way as a pellet. If you use a swimfeeder, fish with pellets and/or hempseed in a 'block-end'. Hook sizes 6 or 8.

Generally always a reserve bait and selected particularly for chub, barbel, bream and dace (and employed by me, back in the day, when eels were a persistent nuisance), sweetcorn can be fished either on the hook or on a hair rig, using grass or a boilie stop to secure one or two pieces. Like meat, corn's adaptability to be used on the hair can save time re-tying rigs whenever a change of bait is called for. If used with a swimfeeder, put hemp in a block-end or ground-bait in an open-ended or cage feeder – hook sizes 6 or 8 for barbel and chub, 10 or 12 for bream, 14 or less for roach, dace and smaller bream.

Maggots are still the first choice with most anglers fishing the leger/'feeder, and while in specimen chub and barbel angling they have been superseded by the ones listed above, they are still normally the bait 'at point of entry' if you are after any of the other species. While cyprinid species (chub, barbel, bream, roach and dace) can be caught on just about any of the generally used coarse fishing baits, more predatory species such as the perch will normally only be taken on leger with maggot, caster, worm or a live or dead minnow. Maggots need to be fished on the hook – a single or double on a size 14 hook or smaller for dace and roach; bunches of as many as you can fit on a size 6 or 8 for barbel and chub; and somewhere in-between for bream and perch. If using a swimfeeder, fill a block-end to the brim with maggots on every cast.

Casters are maggots undergoing metamorphosis – the life stage between larva and adult fly – and can be bought from some tackle shops (you usually need to phone and ask in advance) or you can 'turn' your own by leaving maggots in a tub for a day or two in warm weather. Once you get to the riverbank, fill the tub with water and skim off and discard any casters that float. Casters can be a very good legering bait for all river species and can often score when maggot and others have failed. However, on heavily fished venues, chub and roach in particular are well known for their habit of 'crushing' casters on the hook to extract all the edible contents without giving more than the faintest bite. This means, where the fish are active, you can be left fishing with nothing more than caster skins after only a few seconds! To overcome this, rubber 'false casters' can be bought and placed on the hook along with the real ones, so there'll at least be something left that 'looks' right. If using a swimfeeder, try a block-end filled with hempseed and a few casters.

Worms are an often-neglected bait these days, but being the only truly natural one in common use, it can work when all others fail. A useful bait for upstream legering – if you use worm in conjunction with a swimfeeder rig, chop a few spare ones up, mix into some ground-bait and stuff the mixture into an open-ended or cage feeder. Worms can also be an effective 'cocktail' bait for the larger cyprinid species in combination with sweetcorn, maggot or caster. Hook sizes as for maggots and casters – different sizes of worm are obviously appropriate to whatever size of fish you are after.

Bread will work in summer, tempting any and all cyprinid species on its day; however, this is a far more effective bait for chub and roach fishing in autumn and winter – see Chapter Fourteen for details. The set-up for summer fishing is exactly the same.

A freshly killed minnow trundled along the bottom of a deep pool will catch chub and perch in particular, but it will also sometimes stir the predatory instincts of the barbel. Pike, of course – where they are present – will frequently accept the invitation, as will any large trout in the vicinity. Hook size 6.

* * *

As we have just seen, leger and in particular swimfeeder fishing are all-encompassing methods with regard to modern river coarse angling. Nonetheless, it would be a mistake to overlook float fishing as a means of catching fish from running water. Nowadays, stick float fishing still enjoys a good following, with its main application in match fishing, as well as in pleasure angling where the angler is intent on getting among good numbers of shoal fish like roach and dace. Even so, the stick, along with heavier floats such as the avon and loafer – and even some free-lined baits – can also account for their fair share of larger species, particularly the chub.

Trotting is the most popular form of river float fishing and it's this technique which generally accounts for the largest numbers of roach and dace caught on North East rivers, particularly from moderately paced sections in summer. The idea is to find a run of fairly steadily flowing water of even depth – such a feature going down the near bank being easiest, while it gets steadily trickier the further out into the river the run begins. Finding one which deviates diagonally from near bank to far, looking downstream, would be easiest of all in terms of line-control. The method involves the use of a light stick float fished slightly over-depth and shotted shirt-button style, cast downstream into the run at about 45 degrees and allowed to drift downstream on the current.

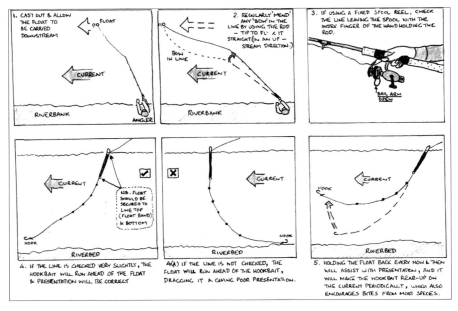

Diagram 23: Trotting – general principles.

To fish in this way correctly, it is necessary to stay in contact with the float by continually checking and mending the line – thus avoiding a 'bow' from developing between the rod tip and float (see Diagram 23). This also ensures that the hook-bait will be travelling close to the riverbed a foot or two ahead of the float. This last part is key to the correct presentation of the bait, as if the line is left unchecked, the float will move ahead of the hook-bait and drag it along at what might appear an unnatural rate to the fish (the water on the surface always travels faster than at the bottom, due to the braking effect of the riverbed).

Trotting is best practised using a centre pin reel, but a fixed spool will still do the job – keep the bail-arm open after casting and check the line by placing the forefinger of the hand holding the rod against the rim of the spool. Striking is achieved by keeping the other hand free to turn the reel handle, thus closing the bale in the same action as the strike.

Pre-baiting is not necessary for trotting, but loose feed should be introduced just ahead of the float on each cast. Maggot is clearly the best bait to use for this technique, although worm, caster and sweetcorn will also work. A standard 11- or 12-foot match or float rod will have the ideal action for this kind of fishing, with the length essential for controlling line on the water at distance. Reel line should typically be 3-lb BS, with a hooklength in the region of 2 lb optional but often preferred by many anglers – attach to the main line by a blood or four-turn water knot. Hook sizes 14 or less.

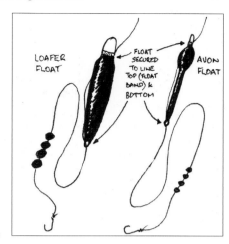

Diagram 24: Loafer float.

While fishing for larger fish like barbel and chub is generally more effective using leger and swimfeeder techniques, float fishing can also work where these species are found in clear runs like that described above. The technique is exactly the same as for 'conventional' trotting, except that, obviously, the rod and reel-line both need to be stepped up and hook sizes should be 6 or 8. The beauty of taking an avon-style rod (as opposed to a 'heavy feeder' or 'quivertip' rod) on a barbel or chub fishing session is that, unlike the other two, the avon is ideally suited to both bottom fishing *and* this heavier form of float fishing. An eleven-foot avon will cope quite ably with the job of controlling line on the water, and combined with a heavier float (an avon or loafer – see Diagram 24) with the shot bulked about a foot from the hook, and 6-lb line, a trotted bait might just score on a day when more static ones have been ignored. Maggot or lobworm are probably still the best baits to try but others can work, particularly corn, although luncheon meat and large pellets will generally be ignored unless they are bumped along the bottom – a type of presentation probably best achieved by a 'trundling' touch-leger set-up.

Last but by no means least is the time-honoured chub fishing technique of 'free-lined slug'. This method is exactly what its name suggests, a large black slug (look for them on the ground beneath trees after rainfall) impaled on a large hook (size 6) cast on a line devoid of any weight or float. This technique is best deployed in a classic tight 'chub hole' with overhanging branches from which the hapless molluscs will regularly fall to be devoured by the leviathans waiting below. If you know of such a place, take a few slugs in a bait box stored with a few freshly picked doc leaves for them to munch on. If, on a blank day, the chub are refusing everything else, this method might well prove your salvation – being a natural bait that most northern chub will rarely ever have been fooled by. Use the same rod and line as for 'heavy' trotting.

CHAPTER THIRTEEN

Autumn: Late Season Nymph Fishing and Cold Weather Trotting for Grayling

Autumn is the time of year when those more hardy game fishermen – the ones that don't stow their tackle away in the cupboard at the first signs of frost – begin to plan ahead for their fishing during the colder months. Thankfully, unlike the situation they had to endure as little as twenty years ago, when almost everything in the North East of England just used to close down, there are now plenty of options open to these keenest of fishermen, whatever their angling preference.

Nowadays, most commercial stillwater game fisheries stay open all winter long, providing their patrons with the chance to fish for stocked rainbow trout in snow, frost, hailstones or sunshine, however there are opportunities for those anglers loath to forsake running water. This chapter looks at the cold weather game fisherman's latter alternative – fly fishing for grayling, which can be practised throughout the autumn given favourable weather conditions; as well as a brief insight into the coarse fishing variation of long-trotting for this species that will often get results right through to mid-March.

The grayling provides a welcome bonus on many North East rivers, with the season for this graceful lady of the stream coinciding with that for coarse fishing and running right through until 14 March. Despite being a member of the salmon family, the grayling spawns in the spring and, like most coarse fish (which many anglers consider the grayling to be), this means they're in prime angling condition come the autumn and winter. What's more, the grayling can often be relied upon more than any other species to feed in even the coldest of water temperatures. This means that those hardiest of fishermen can still pursue them when there's snow on the ground or in the middle of one of those hideous week-long spells of freezing fog that keep air temperatures below zero for days!

Huge alterations in approach or tackle choice are not necessary, either – this being a most accommodating species to fish for. Most of the techniques that worked for river trout in the spring and summer will remain effective for winter grayling fishing – with subtle variations – as will the coarse fisherman's

An autumn grayling.

staple of float-fished worm or maggot. Besides the closest Yorkshire rivers (the Swale and Ure both contain them), in the North East the grayling is found in the Tees, Wear, Derwent, Blyth, Till and Tweed, with clubs and riparian owners on most of these rivers promoting grayling fishing as a sport for at least part of the winter.

Despite its aforementioned seasonal unity with coarse species, in anatomical and behavioural terms, the grayling is very much a game fish, meaning that its diet and habitat preferences are extremely similar to those of the brown trout in everything other than that it is characteristically a shoal fish. It follows therefore that grayling spend much of their lives chasing aquatic insects in all their forms, and on North East rivers there is no more rewarding nor effective method for colder weather than the Czech nymph, a modern variation of the more traditional 'upstream nymph' technique of fly fishing. Czech nymphing was really designed for use on larger rivers, such as the Tees or Wear, where an angler would ordinarily be waded in, casting a heavily weighted nymph on a fairly short line a few yards slightly upstream, allowing it to drift down as the fly sinks to a point about 45 degrees downstream. However, on smaller rivers such as the Derwent, this technique can be employed from the bank (or at least from a point fairly close to it), hence there's rarely any need to wade deep. Whether waded in or not, the key to this method is to retrieve line and stay in contact with the nymph, watching for the movement on the end of the fly line that indicates a taking fish, with this almost invariably being

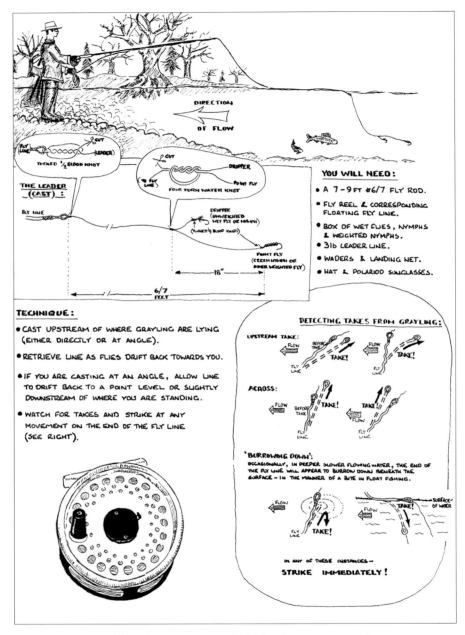

Diagram 25: Principles of upstream nymph fishing for autumn grayling.

seen rather than felt. A fly rod of 8 or 9 feet combined with a number 6 or 7 floating line will suffice for almost all situations on Northumbrian rivers (see Diagram 25 for the general outline).

Beginning with a cast of say ten yards or less, made either directly or obliquely upstream, in upstream nymph fishing the line is allowed to drift back towards the angler while recovering and watching it carefully for telltale pulls against or across the current. The leader you use needn't be any longer than about six feet of 3-lb line, the weighted nymph being the point fly and any dropper (a second wet fly or nymph can also be added thus) being tied on about eighteen inches higher up. Depending on the weight of the nymph used, the point fly will cause the business end of the rig to sink in proportion to the strength of the current, while the point at which the leader joins what has to be a floating fly line – sinking lines won't work for this – will remain on or close to the surface. A stronger current will demand a heavier nymph, as grayling are nearly always found near the bottom, and a light fly will cause the whole thing to drift ineffectively over their heads. Conversely, too heavy a nymph in too light a current will cause the point fly to sink too quickly, with the infuriating result that you'll eventually find your 'take on every cast' is the bottom and you'll have to pull for a break. Like with all good fishing, correct balance is the key.

While Czech nymphs and various heavily weighted shrimp patterns are frequently preferred for grayling fishing in the higher-water conditions often encountered late in the year, in lighter flows traditional weighted patterns like the GBHE and Pheasant Tail can be just as effective, if not more so. Just remember to always make your cast a fair way upstream, as in my experience North East grayling won't take any fly that has started to swing round in the 'down and across' trout wet-fly mode. If you think about it, it stands to reason why not: at the point at which the cast passes 90 degrees, the combination of pressure from the rod tip and the force of the current start to pull the point fly up and away from the riverbed on a tight line – and up and away from where the grayling are lying. If therefore, for whatever reason, you think you're fishing upstream of a shoal of grayling, don't be tempted to fish for them from upstream if you can avoid it – just move downstream and cast up.

When looking for a few of the best kinds of place to fly fish for grayling look for deeper, swiftly flowing water and areas of uneven current – creases, rock sills or weirs where fast water gushes into a pool – which is always a good start when trying to pinpoint this species. Such places abound on most northern rivers, which aren't termed 'trout streams' for nothing – just remember the simple rule that your fly needs to be close to the riverbed. Something else to watch out for in this and most of the other places you will encounter is overhanging branches, which by making false casts low over the water, or by carefully roll-casting your line back upstream each time, are a hazard that is easily avoided.

It goes without saying that weir pools are likely to be hot spots on any river where grayling are present – the current, especially at this time of year, providing the ideal habitat for what is in effect the game fishing equivalent of the barbel. The general rules already described also go for weir pools, although due consideration should be given to the likely undertow you'll experience the closer you fish to the weir sill. Where wading is necessary, while wellies will often suffice, a pair of thigh waders would allow you to get closer to many of the better lies. Another word of warning here, however, as a pair of polaroids might be a worthwhile investment – if nothing else, then just to help you see more clearly exactly where the shallow water shelves off very suddenly – a regular aspect of such angling features. The rest is fairly straightforward – cast a few yards upstream – stay in contact with the nymph and be prepared for palpitations (and to strike!) whenever the line suddenly whizzes off very purposefully either upstream or across.

Very late in the season – from the turn of the new year onwards – the temperature of the water, combined with the force of the current, can make fly fishing much less of a practical consideration. This is where bait fishing, and those techniques more normally associated with coarse angling, come into their own. Of course, for many years the grayling was actually considered a coarse fish – indeed it was removed from many rivers by game anglers who considered it a pest, as it competed with the trout for food and habitat. Nowadays, thankfully, a more sensible approach is in vogue and, as well being as a species that can lengthen the river game angler's season, it can also be seen as a fish to be sought out by coarse anglers when the water temperature goes so low it puts all the others off.

Grayling will often still feed in river water so cold that the ice on its surface in slack areas has first to be broken before fishing can commence – conditions not unknown in the North East from late December through to mid-March. While the chub is a reliable adversary on many northern rivers in all but the very coldest weather, in freezing temperatures it is less likely to feature – best wait for a thaw, and the swollen river conditions brought about by mountain snow-melt. These same temperature parameters also apply to the roach and, to some extent, even the perch.

However, the often static approach favoured by the winter chub angler is not appropriate to ultra-cold-water grayling fishing. In such conditions, the shoals of grayling are likely to be tight and could be just about anywhere in the river, so a roving approach is always best – and trotting, in the manner described in detail in the last chapter (summer) is the only way to effectively probe the most likely runs.

While the technique you employ is exactly the same (except of course for the probable greater speed of the current), the baits you can use are more limited than those outlined for general summer coarse fishing. Like all members of the salmon family, the grayling is predatory, so while baits such as bread and corn do tempt the occasional specimen, 'live' baits like worm or maggot are far better. Worms should usually be on the small side – redworms, brandlings or denrobaenas all being about the right size, for while grayling have smaller mouths than trout or chub, they are also wont to pick at the tail of a long worm, such as a lob – giving fast bites without ever getting hooked. Incidentally, the worm should only be hooked once – close to the head.

Maggots afford the winter grayling angler the advantage of being able to introduce (sparingly) loose-fed maggots with the occasional cast, which should, as closely as possible, follow the trajectory of the float-fished hook-bait. In low temperatures, all fish – even grayling – have a tendency to become picky, so choosing a particular colour of maggot and sticking to it can sometimes be the key to success, if you have first to earn the fish's 'trust'. Trial and error is the only real way to discover which colour is most effective, but choosing natural (white) maggots would always be a good place to start.

Of course, not all angling clubs allow bait fishing, even in winter, so do make a point of checking the rules before you venture out. However you fish for them, the autumn or winter grayling can make for an excellent extension to any angler's season.

CHAPTER FOURTEEN

Winter: Spate River Coarse Fishing for Chub

If, like me, you regard the summer months as a fleeting window of opportunity to fish for that hardest-fighting of river fish, the barbel, you'll be all too well aware that up here in the far north of England the nominally lengthy season for this species is curtailed by cold weather a good six months before the local byelaws kick in. Despite reassurances given in textbooks and articles written by southern-based scribes, rising barometric pressure and its stimulating effect on the winter barbel's appetite does not hold true for rivers in this part of the world. The plain fact is that that almost all the rivers up here rise on high moorland and, as such, water temperatures from late September onwards are rarely comparable to those in low-lying southerly flows.

Up here, therefore, any river angler in search of big fish sport needs to look to another species once the weather turns, and fortunately for us the gap is suitably filled by one that continues to feed right through the winter. There won't be many northern barbel anglers that haven't caught a chub at some point in their career, given its presence in most of the larger northern rivers. Indeed, most will probably have caught at least as many as they have barbel, owing to the chub's obliging acceptance of most baits used for its larger relative.

Nonetheless, I'm willing to bet that most barbel anglers that have caught summer chub – even big ones – probably regarded their capture as decidedly 'second best', which while not strictly a case of angling elitism, has much to do with the latter species' apparent lethargy in warmer weather. While the summer chub is sadly misunderstood – biologically at least – it is a fact that pound for pound they don't come close to their distant cousins as a sporting fish. And, although a big one can momentarily have you thinking 'barbel' as a flick of their powerful tail takes line against the reel clutch, the deception rarely lasts. Most fair weather chub come to the landing net sedately to say the least, but an angler that has yet to experience a winter chub might be in for quite a surprise if they ever hook one on a flooding spate river when there's a nip in the air!

Winter chub fishing on the River Wear.

The winter chub is almost like a different species compared to its warm weather doppelganger. While the summer-caught fish lacks condition owing to the rigours of springtime spawning, in winter it is fully recovered and able to put pressure on the angler's tackle in full proportion to its often considerable weight. On northern rivers where the chub is found (those to the north of the Tyne don't contain them) it is typically one of the largest non-migratory species, lagging behind only the pike and the barbel in terms of average weight, but out-sizing all the rest bar the very occasional exceptionally big brown trout. On the rivers Wear, Tees, Swale and Ure, while juvenile specimens (often mistaken for dace) and fish in the 1½- to 3-lb range are regularly taken by anglers looking to 'bag up', upwards of this size they tend to be the preserve of the specialist chub and barbel angler. Four- and five-pounders, while considered to be specimen fish, are far from uncommon on any of these rivers and even six-pounders are not unheard of, although the Tyne's chub, possibly due to the colder average water temperatures caused by Kielder Reservoir discharges, do not seem to reach these sizes.

In these parts, the chub season 'proper' can be said to start at about the same time as the barbel start to lose interest, which is typically from the about third week in September through to mid-October, depending on how warm the summer has been. In a normal year, the drop in water temperature on the river will coincide with the first regular spates, and these factors put together will prompt the chub to seek out deeper water with a steadier current, largely abandoning those shallower, swifter runs where they're often caught in the summer. There is a period of transition of course, and there are also those places where the water is deep yet the oxygenation sufficient for the chub to inhabit all year round. Places close to where glides or rapids break into deeper pools or sharp turns where deep water gathers pace on the outside of a bend are obvious examples.

And familiarity with such hot spots can give the winter angler a distinct advantage, as knowledge or at least a good understanding of the chub's exact whereabouts is often the key to success in colder weather. 'Chuck it and chance it' is not really an option in winter, unless you're willing to spend uncomfortable hours on the bank, bite-less! In colder water temperatures, fish are rarely willing to expend energy searching the river for food. Nature and the river current mean that they can just stay in one place and wait for it to come to them. Introducing loose-feed in such circumstances *will* draw fish to your hook-bait – and this is an important aspect of winter chub fishing – but it only works if you're casting to within a few yards of where the fish are lying.

So the principles are clear-cut, but what about putting them into practice. Here in the North East, our best coarse fishing rivers are confined to the south of the region, and in their lower reaches, the rivers Wear and Tees are prime examples of good chub rivers. Down in Yorkshire, almost all of the main rivers are good for the chub, with our closest two, the Swale and the Ure arguably the best of the bunch, upstream of their confluence near Boroughbridge. One thing they all have in common is that chub fishing comes into its own once the leaves are all down off the trees and morning frosts are pretty much an everyday occurrence, but being spate rivers, the fishing in winter is never going to be as reliable as it is in the warmer months. While the chub might well provide better sport at this time of year, the quirks of the winter weather – and in particular that on the higher ground where the river rises – will frequently affect the ability of an angler to access the river and its fish.

One of the most frustrating things the winter coarse angler will experience is arriving to find the river in flood, several feet above even normal winter levels and with the current rushing by in the guise of a muddy chocolate fountain. On larger rivers, this will normally follow a period of prolonged heavy rain or a mild spell precipitating the thaw of a fall of heavy snow on the hills. But will this necessarily make the river completely un-fishable?

You would think so – especially in the likely event that the flood water will probably consist, at least partly, of snow-melt off the mountain tops, with the probable outcome that this will reduce the water temperature to only a few degrees above zero. And what about the current? Even if you can get onto a safe part of the bank to fish, the river may be running at such a pace that even a six ounce grip lead isn't going to hold bottom. Just about every other species will go off the feed in such conditions, but find the right place and the good old chub can still relied upon!

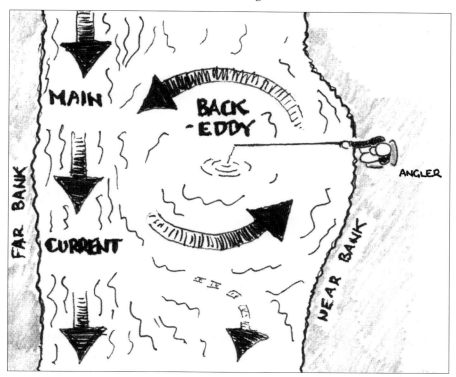

Diagram 26: Fishing a slack.

This is where that 'familiarity with a river' I mentioned earlier can come into its own. While most stretches will indeed remain completely un-fishable until the water levels subside, there is one particular type of feature that can be fished in all but the very worst of spates. Slacks (Diagram 26) are back-eddies formed by depressions in the riverbed where the current (characteristically on one side of the river) flows back round on itself to produce a rotating body of water fed by the force of the current running down the other side and into 'the hole'. If you can safely get access to the bank on the near side of the slack, you're in what is potentially one of the best winter chub fishing situations you can get.

The key to this kind of feature's success is that the current near the middle of the river is normally quite sedate, even though there will characteristically be the fieriest of spates bombing down the far side. Nevertheless, your tackle choice should still be in proportion to the general river conditions (and the potential size of the quarry) – with a stout avon-style or heavy feeder rod combined with a main line of 5–6 lb and hook length not less than 4 lb.

The only technique that will really work in this situation is the swimfeeder, although the nature of a slack means that choosing a feeder weighing more

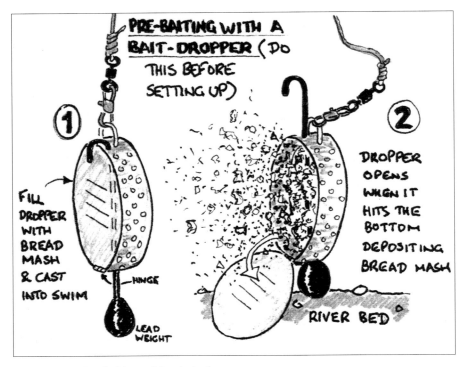

Diagram 27: Pre-baiting with a bait-dropper.

than a couple ounces is generally unnecessary. Most of the baits normally used for winter chub fishing will work (and that means just about anything!) but to get the best out of this sort of swim in a winter flood you'll need to 'use your loaf'. Oft-neglected and sometimes feared by many anglers (to be honest, I never believed this bait would stay on the hook in a swirling river either) the use of bread flake is a devastatingly effective technique – as long as you keep fairly tightly to the advice that follows.

The first thing you need to do is prepare a large bag of breadcrumbs the night before fishing by crushing a large loaf of fresh white bread in a liquidiser. Take the breadcrumbs to the riverbank in a bag and mix them with river water in a bait tub, in the same way as you would with ground-bait. The bread-mash can then be bowled into the swim in balls before you start to set up your tackle, or in the case of venues that enforce ground-baiting restrictions, introduced in a bait dropper (see Diagram 27).

The feeder set up is as standard, but a cage feeder is used rather than a block-end or open-ended feeder, as this is the best type for feeding with bread mash (Diagram 28). The bread you use as hook-bait must also be fresh – buy it on the morning of your fishing trip and while sliced bread is easier, if you can, get bakery fresh rather than Sunblest/Warburtons, etc. The flake is made

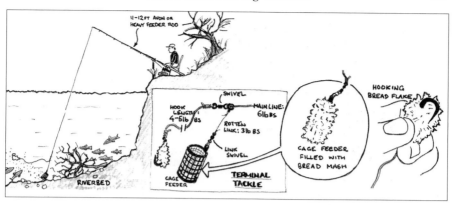

Diagram 28: Feeder fishing in a slack.

by tearing a thumb-sized piece of bread from a slice and simply wrapping it round the shank of a size 6 hook, leaving the hook point exposed. Bread paste can also be used thus – this is made by mixing bread or flour into a doughy consistency with water, and it will stay on the hook more assuredly – but it's unlikely to be as attractive to the chub as flake. Mould a fistful of bread mash into the feeder, place your cast smoothly into the centre of the slack (jerky casting is more likely to make the bait fly off) and wait for bites – if you've pre-baited they might come sooner than you think.

Bites will usually be positive or not at all and once you've hooked a chub in this kind of place you'll soon see why I place such an emphasis on the use of stout tackle. The nature of any slack means there'll be all manner of debris lying unseen six to ten feet down on the riverbed – tree branches and all, deposited by successive floods – and a hooked chub will make every effort to get to them. Keep the fish's head up and, providing you've not cast right into one of the snags (trial and error is the only way to find out where they are, so bring plenty of feeders and use a rotten link to attach them) even the biggest chub shouldn't be able to dictate the fight. Keep your wits about you until the fish is safely in the net and don't be under any illusion that it's going to just give up in the way a summer chub might do!

Of course, the rivers won't be in flood all winter and cold gin-clear water can present its own set of problems to the winter chub fisherman. The slacks you might favour during high water still swirl round when the river returns to its normal winter level and chub will still respond to similar tactics as in flood conditions, but other spots, and baits and techniques, can also be brought into play when the river level drops.

Some of these will fall into that aforementioned category of 'places where the water is deep, yet the oxygenation sufficient for the chub to inhabit all year round' and as far as baits and tactics go, you can take your pick here.

A maggot or worm trotted through the deepest part of the river will take chub; but this is equally or probably even more likely to pick up the dace, grayling and small trout that come back into play when the river level drops. Nevertheless, as I said in the last chapter, a bait moving through and probing the river to find the fish (rather than waiting for them to find you) can be a better approach in cold water conditions, so when attempting the more effective chub tactic of legering, it can pay to bait up a couple of swims first and search out each one in turn with a trundling leger set up until you get signs of interest – there's unlikely to be much competition for space from other anglers on a cold winter's day!

Winter fishing is generally seen as a specialist pursuit and understandably so. However, with the extra comfort provided by the correct kind of cold-weather clothing, plus a bit of attention paid to the kind of detail just described, a decent day's fishing in the colder months can be a worthwhile experience for any angler.

APPENDIX ONE

Stillwater Coarse and Game Fishing Venues in Northumberland and Tyne & Wear

NORTHUMBERLAND (COARSE FISHING)

Bolam Lake is a decent sized lake of around 25 acres, situated about 2 miles off the A696 at Belsay – follow brown road signs. *Species:* pike, perch (and roach and rudd if you know where to find them) *Permits:* Belsay post office – phone 01661 881207. *Website:* www.northumberlandlife.org/bolamlakecountrypark/AboutUs.asp

Brenkley Pond (near Dinnington village) is close to the Milkhope Centre – from the A19/A1 Seaton Burn Roundabout, take road past the Holiday Inn, then the first left turn. *Species:* carp, tench and other mixed coarse. *Permits:* yearly membership available from Wansbeck & District Angling Club. *Website:* www.wacac.me.uk/

Dissington Pond is just off the A696 about a mile the Belsay side of Ponteland village. *Species:* mixed coarse, tench, carp. *Permits:* no day tickets, you have to be a member of Big Waters AC. *Website:* www.bigwatersanglingclub.dreamstation.com/

Felton Fence Farm Fishery. Take the A1 to Felton, or the A697 to Lonframlington. Felton Fence is just off the B6345, which connects these two villages. There are several ponds. *Species:* carp, tench, perch, roach, rudd. *Permits:* day tickets available on-site. *Website:* www.feltonfence.co.uk

Horton Grange Lake (near Dinnington village). From the A19/A1 Seaton Burn roundabout, take road past the Holiday Inn, then the second left turn. *Species:* roach, perch, bream, tench, ide. *Permits:* yearly membership available from Wansbeck & District Angling Club. *Website:* www.wacac.me.uk/

Milkhope Lake (near Dinnington village). At the Milkhope Centre – from the A19/A1 Seaton Burn roundabout, take road past the Holiday Inn, then the second left turn. *Species:* carp, tench, ide. *Permits:* yearly membership available from Wansbeck & District Angling Club. *Website:* www.wacac.me.uk/

QE2 Lake. A fairly big lake (40 acres) near Ashington, the main car park is that for the QE2 County Park, beside the A189 'spine road' between Woodhorn & Ellington. *Species:* pike, carp, bream, roach, rudd. *Permits:* available on-site from the Lakeside Hotel, next to the car park, or from McDermott's tackle shop, 112 Station Road, Ashington. *Website:* www.wacac.me.uk/

Whittle Dene is a complex of small to medium sized reservoirs straddling the junction of the B6318 'military road' and the B6309 about 10 miles west of Newcastle – to get there, go to Heddon-on-the-Wall and take the B6318. *Species:* dace, perch, roach, gudgeon, bream. *Permits:* day tickets available on-site. *Website:* www.nwl.co.uk/Gofishing.aspx

TYNE & WEAR (COARSE FISHING)

Angel of the North Lakes is between Birtley and Eighton Banks on the B1288. Can be accessed from the south at the first junction after J65 on the A194(M) or, going south, at the last exit from the A1 (A1234 Washington) before J65. *Species:* carp, tench, bream, rudd and other coarse. *Permits:* available on-site. *Website:* www.angelnorthlakes.com/

Big Waters (22-acre lake) and **Little Big Waters** (adjoining pond) are accessible from the Big Waters Nature Reserve car park at Brunwick Village, near Wide Open. *Species:* carp, bream, perch, roach, rudd. *Permits:* no day tickets, you have to be a member of Big Waters AC. *Website:* www.bigwatersanglingclub. dreamstation.com/

Killingworth Lakes, North Tyneside, is situated either side of Southgate in Killingworth, with a large and a small lake, divided by the road. *Species:* carp, tench, roach, rudd, perch, crucians and pike. *Permits:* day tickets available from Killingworth Leisure Centre, with disabled access all along the southern bank of the large lake. Controlled by the Tyne Anglers Alliance. *Website:* www.thetaa.co.uk/menu.html

Leazes Park Lake, Newcastle City Centre (between the RVI and St James' Park). *Species:* carp, tench, bream, perch, roach. *Permits:* day tickets available from local tackle shops (nearest is Bagnall & Kirkwood, Grey Street) or refreshment hut at weekends. Leazes Park AA. *Website:* www.leazesangling. com/

Marden Quarry, Whitley Bay, is in the park alongside Studley Gardens and the Broadway (A193). *Species:* carp, raoch, rudd, perch, bream, crucians and the occasional large goldfish! *Permits:* day tickets are available locally from Frasers Angling, ID Fishing and Billy's of North Shields. The lake is a Big Waters AC water. *Website:* www.bigwatersanglingclub.dreamstation.com/

Silksworth Lakes, south of Silksworth Sports Complex in Sunderland. *Species:* carp, bream, tench, roach, rudd. *Permits:* phone 0191 522 9070. Sunderland Fresh Water AC.

Throckley Reigh is half a mile north of the tidal River Tyne in the Tyne Riverside Park at Newburn. *Species:* mixed coarse. *Permits:* controlled by the Tyne Anglers Alliance, to which several NE angling clubs are affiliated. *Website:* www.thetaa.co.uk/menu.html.

NORTHUMBERLAND (GAME FISHING)

Fontburn Reservoir is Northumbrian Water's original big fish trout Mecca and is situated about a mile off the B6342 roughly halfway between Rothbuty and its junction with the A696 Newcastle–Jedburgh road. *Species:* brown trout, rainbow trout, brook trout. *Permits:* available on-site. Multi-bait fishing allowed. *Website:* www.nwl.co.uk/Gofishing.aspx

Hallington Reservoirs (there are two separated by a causeway) are controlled by Westwater Angling Club and there are only a limited number of day tickets available. The reservoirs are a third of a mile from the village of Colwell on the B6342 (which connects Hexham and Rothbury as well as both the A68 and A696 Jedburgh roads). Species: brown trout, rainbow trout. Permits: contact Westwater Angling on 01434 681405 or info@westwaterangling. co.uk. *Website:* www.westwaterangling.co.uk

Hazon Burn Lakes is just south of Alnwick off the A1. *Species:* brown trout, rainbow trout. *Permits:* available on-site. *Contact:* 01665 881000.

Higham Lakes Trout Fishery is 2 miles north of Ponteland on the A696. *Species:* brown trout, rainbow trout, blue trout. *Permits:* available on site. *Contact:* 07850 428195.

Kielder Water is the largest man-made lake in Western Europe and is a key part of Northumbrian Waters trout fishery emporium. The reservoir is situated 10 miles west of Bellingham (which is found by taking the B6320 off the B6318 'Military Road' at Chollerford) in the upper North Tyne valley. Follow the brown signs. *Species:* brown trout, rainbow trout. *Permits:* on-site, boat hire also available. Multi-bait fishing allowed from the shore. *Website:* www.nwl.co.uk/Gofishing.aspx

Langley Dam is only a few miles west of Hexham close to the A686 Haydon Bridge–Alston road. *Species:* brown trout, rainbow trout. *Permits:* available on-site. *Contact:* 01434 688846.

Sweethope Loughs consists of two lakes about four miles west of the A696 Newcastle–Jedburgh road near the village of Kirkwhelpington. The larger lake is a vast 125 acres (half as big again as nearby Fontburn Reservoir) and boat hire is available to cover this lough's vast surface area. The smaller lake is a 'mere' 24 acres and fishing is from the bank only. *Species:* brown trout, rainbow trout, blue trout. *Permits:* available on-site. *Website:* www.sweethope.co.uk

APPENDIX TWO

Coarse and Game Fishing on Rivers in Northumberland and Tyne & Wear

River Aln. The Aln Angling Association has 5 miles of salmon, sea trout and brown trout fishing between Alnwick and Alnmouth. Permits can be obtained from Hardy & Greys Ltd, Willowburn Industrial Estate, Alnwick, tel: 01665 510027, or Jobsons of Alnwick, Tower Showrooms, tel: 01665 602135.

River Blyth. The Bedlington & Blagdon Angling Association controls some 6.8 miles of mostly double-bank fishing on the River Blyth for brown trout and grayling. The association's water stretches from the bridge on the A1 at Stannington through Plessey Woods Country Park and the Hartford Hall Estate to the stepping stones at Humford near Bedlington. *Website:* www. babaa.org.uk/

River Coquet. The Northumbrian Anglers' Federation was founded in 1894 and first took the lease for much of the fishing on the Coquet in 1897. It still to this day controls over 10 miles of river between Thropton, at the bottom of Coquetdale, and the estuary between Warkworth and Amble, with the principal beats being from Rothbury to Pauperhaugh, Felton to Sandy Hole and the 'Salmon Only' tidal stretch at Warkworth. Permits can be bought for either Brown Trout Fishing Only or Salmon and Sea Trout Fishing. Weekly or day visitor permits are also available. Website: www. northumbriananglersfed.co.uk/

River Derwent. Forming the boundary between Northumberland and County Durham for most of its length, the last 8 miles of the Derwent flows through the rural fringe of Gateshead Metropolitan Borough and the Axwell Park & Derwent Valley AA controls most of it. Fishing is for brown trout and grayling and season permits are available directly from the club, with day tickets also sold at Rowlands Gill Caravan Site Shop. Website: www. apdvaa.co.uk/

River Rede. A tributary of the River North Tyne, this small upland stream rises near the Scottish border. There are two stretches open to public angling for Salmon, Sea Trout and Brown Trout. Otterburn Tower is controlled by the Castle Hotel in Otterburn village (on the A696 Newcastle–Jedburgh road), contact Mr John Goodfellow on 01830 520620 or visit: www.otterburntower. com. Permits can be obtained from the hotel reception. Woolaw Farm is 6 miles north of Otterburn on the main Newcastle–Jedburgh road after it has become the A68. Permits are available from Woolaw Farm, by telephoning 01830 520686 or visiting the website: www.woolawfarm.co.uk

River Tyne. The largest river system in the North East, there are a number of locations within the Tyne catchment that can be fished with the Tyne Angling Passport, a day ticket arrangement between a number of local angling clubs and the Tyne Rivers Trust (see links). Tyne Angling Passport Webpage: www. tyneriverstrust.org/index.php/home/what-we-do/conservation/tyne-angling-passport

The River Tyne at Tyne Green, Hexham has a mile of fishing available on the south bank, upstream of Hexham Bridge (A6079), adjacent to Tyne Green Country Park. The river here contains salmon, sea trout, brown trout, chub and dace. Day, weekly and season permits are available from Hexham Tourist Information Centre, Wentworth Car Park, Hexham. Tel. 01434 652220.

While the Tyne is predominantly a game fishing river, it does naturally contain a number of coarse species. The Tyne Anglers Alliance, to which a number of local angling clubs are affiliated, controls a section of the north bank of the tidal River Tyne in Tyne Riverside Country Park at Newburn. This part of the river holds brown trout, dace, roach and eels, as well as – in keeping with its tidal nature – the occasional flounder. *Website:* www.thetaa. co.uk/menu.html

Others fishing on the main River Tyne include the Wylam Angling Club, which controls fishing at Wylam and Hagg Bank – information and day tickets are available from the Spar shop in Wylam village or by phoning 01661 852214. And the Northumbrian Anglers' Federation controls two sections near Wylam and Ovington – contact details/website are the same as those for the River Coquet.

River North Tyne. The northern upstream arm of the main Tyne, the North Tyne has fishing for salmon, sea trout and brown trout available about 4 miles from Bellingham on the B6320. The contact for permits (bookings in advance) is by telephoning 01434 240239.

River South Tyne. The Haltwhistle Angling Club have 6 miles of the River South Tyne available for salmon fishing at £20 per day, £50 a week or £120

for a season ticket. Permits are available from the club website: http://www.haltwhistleangling.co.uk/. The Alston & District Angling Club (just over the border into Cumbria) controls fishing for salmon, sea trout and brown trout from Alston to Langley Viaduct. Permits can be obtained from Alston post office and Tourist Office.

River Till. The Lower Tindall and Red Scar beats offer salmon, sea trout, brown trout and grayling fishing. *Contact Address:* Redscar Cottage, Millfield, Wooler, Northumberland, NE71 6JQ, telephone 01668 216223. *Unlike the other rivers in the area, the Till – a tributary of the Tweed – comes under the jurisdiction of the River Tweed Commissioners, rather than the Environment Agency, therefore different rules and seasons can apply.

River Wansbeck. The Wansbeck Angling Association controls around eight miles of fishing for brown trout on the River Wansbeck from just upstream of Morpeth to the tidal limit at Sheepwash. Season permits and day tickets are available from McDermotts tackle shop, 112 Station Road, Ashington, Tel. 01670 812214, and Game Fishing Supplies, 3 Fawcett's Yard, Morpeth (near the new bus station), Tel. 01670 510996.

See also Appendix Five (Free Fishing Venues) for information on free fishing on the River Wansbeck at Morpeth.

Stillwater Coarse and Game Fishing Venues in County Durham and North Yorkshire

STILLWATER COARSE FISHING VENUES IN COUNTY DURHAM

Aldin Grange Lakes is near Neville's Cross, in Durham City. Turn off the A167 near the Pot & Glass and head towards Bear Park. *Species:* carp and tench. *Permits:* day tickets available on site. *Website:* www.aldingrangelakes.co.uk/

Brasside Pond Complex. Opposite Frankland Prison, Brasside, Durham City. *Directions:* Take the road off the A167 past the Arnison Centre and continue on down the hill through Brasside village and past the prison to your right. The entrance to the complex is first on the left *after* the turn off to Finchale Abbey. *Species:* carp, tench and pike in the specimen lake. Mixed coarse species in the others. No day tickets: you have to be a member of Durham City AC. *Website:* www.durhamanglers.co.uk/brassidecomplex.html

Eden Meadows Fishery is located off the B1280 near Trimdon Colliery. *Species:* carp and mixed coarse. *Permits:* day tickets available on-site. *Website:* www.edenmeadows.co.uk

Fieldson's Pond is on the outskirts of Shildon near the industrial estate. *Permits:* day tickets from the cottage on the farm. *Species:* mixed coarse. *Contact:* telephone 01388 775044.

Hartlepool Reservoirs. Hartlepool Water (a division of Anglian Water) own three reservoir sites within five miles of Hartlepool. At the Upper and Lower Hart Reservoirs, and the Crookfoot Reservoir, nearby, coarse fishing is managed by Hartlepool and District Angling Club and Hurworth Burn Angling Club. Membership is available from local fishing tackle shops. There are stocks of trout and coarse species. *Website:* www.hartlepoolwater.co.uk/leisure/

Shafto's Lake is situated in Whitworth Park, Spennymoor and controlled by Ferryhill & District Angling Club. *Permits:* day tickets available from the Shafto Inn. *Species:* mixed coarse. FDAC Unofficial Members' *Website:* www. ferryhillanglers.co.uk/

West Farm Lake is situated at West Farm, Old Stillington, about 2 miles from Thorpe Thewles (which is on the A177 between Stockton and Sedgefield). There are carp, tench, rudd, perch and roach. Telephone 01740 631045 for more details.

STILLWATER GAME FISHING VENUES IN COUNTY DURHAM:

Aldin Grange Lakes is near Neville's Cross, in Durham City. Turn off the A167 near the Pot & Glass and head towards Bear Park. *Species:* rainbow, blue and brown trout. *Permits:* available on-site. *Website:* http://www.aldingrangelakes. co.uk/

Balderhead Reservoir, near Barnard Castle in Teesdale, is the first of Northumbrian Water's wild trout fisheries. Leave the B6277 at Romaldkirk or the B6276 near the hamlet of Grassholme. *Species:* brown trout (wild). *Permits:* pay at Grassholme Reservoir (see below). *Website:* www.nwl.co.uk/ Gofishing.aspx

Cow Green Reservoir is another of Northumbrian Water's wild trout fisheries, only this one, at a vast 77 acres, is considered to be one of the most challenging fishing experiences in the North of England. Cow Green is at the head of Teesdale about 15 miles west of Middleton-in-Teesdale. Follow the B6277 in the direction of Alston as far as Langdon Beck and then bear left. The reservoir is four miles along the minor road. *Permits:* honesty box available on-site or pay at Grassholme Reservoir (see below). *Website:* www.nwl.co.uk/Gofishing.aspx

Derwent Reservoir is Northumbrian Water's premier commercial trout fishery. There is something for everyone here, the 1,000-acre reservoir having miles of bank space on either side and including access for the disabled. The more pastoral northern shore is in Northumberland and reserved for fly fishing only, to allow the roving angler a chance to explore the reservoir unimpeded. On the south shore (County Durham), multi-bait fishing is allowed with very few restrictions (spinning is also permitted) although checking the rules first is always advisable. The fishing lodge/shop for Derwent Reservoir is situated right on the B6278 at the foot of the dam, about 6 miles west of Shotley Bridge and just 2 miles from where it crosses the A68 at Carterway Heads.

The south shore, close to the dam, can be found by going up the access road which leaves the B6278 half a mile further along from the lodge, while the other popular spots, Pow Hill (a country park), and Hunter House are further along the B6303 road to Blanchland (continue along the B6278 to Edmundbuyers and bear right, looking out for turnings on the right in two and three miles respectively). The fly-only north shore is found by going up the water board access road to the left (looking from the lodge) to the top of the dam and parking on that side. The whole of that shore is conducive to fly tactics at various times of the year, so it pays to be mobile. *Species:* brown trout, blue trout, rainbow trout. (There are also roach in the reservoir, which will sometimes be caught on 'scaled down' multi-bait tactics.) Permits, tackle and bait supplies are available on-site. *Website:* www.nwl.co.uk/Gofishing. aspx

Grassholme Reservoir covers 140 acres and is one of Northumbrian Water's longest established put-and-take reservoir fisheries, located high up in Teesdale about 6 miles from Barnard Castle. Fishing at Grassholme comes under the same rules and permit prices as those for Derwent, except that there is no fly-only area. There is access for disabled anglers. The reservoir can be found by taking the B6277 from Barnard Castle (after crossing the River Tees, go straight ahead *instead* of turning left onto the A67). Carry on through Cotherstone, before turning left towards the reservoir at the village of Mickleton. It is the most popular of the 'Teesdale' fisheries (it's really on the River Lune, a tributary of the Tees) and is very close to Selset, Balderhead and Hury. *Permits:* available on-site at the Visitor Centre. *Species:* rainbow trout, brown trout. *Permits:* available on-site. *Website:* www.nwl.co.uk/Gofishing. aspx

Hury Reservoir is the only completely fly-only reservoir controlled by Northumbrian Water and is stocked weekly with trout from the company's own Teesdale hatchery. It is suitable for disabled anglers. Directions as for Balderhead Reservoir: Selset is the lowest of the three reservoirs on the River Balder (Balderhead is uppermost), so the Romalkirk turning is closest. *Species:* rainbow trout, brown trout. *Permits:* honesty box on-site or pay at Grassholme. *Website:* www.nwl.co.uk/Exclusiveflyfisheries.aspx

Jubilee Lakes is a highly regarded trout fishery (voted No. 6 in the top fifty UK trout fisheries by *Trout Fisherman* magazine in 2005) situated just off the A68 between Darlington and Bishop Auckland. There are two lakes, 2.5 and 1.5 acres in size, stocked daily with rainbow trout. *Website:* www.jubileelakes. co.uk/

Knitsley Mill Trout Fishery situated at Delves, near Consett, is a 7-acre complex of three lakes set in woodland. The lakes are spring-fed, making the water very clear. *Species:* rainbow trout. *Permits:* day tickets available on-site. *Contact:* telephone 01207 581642.

Selset Reservoir is a wild brown trout fishery close to Grassholme Reservoir, controlled by Northumbrian Water. *Permits:* pay at Grassholme Reservoir (see above). *Website:* www.nwl.co.uk/Gofishing.aspx

Tunstall Reservoir is part-leased by the Ferryhill & District Angling Club and run along similar lines to Waskerley Reservoir (see below). Information on membership and/or day tickets is available by telephoning 07825 951525. Tunstall is four miles north of Wolsingham, which is on the A689 in Weardale. Take Leazes Lane, the second last turning on the right as you go west out of Wolsingham, and follow this road up the valley until you go past the dam.

Waskerley Reservoir, near Consett, offers up to five-day tickets for worm and fly fishing (telephone 0191 488 4873 for details). This reservoir is one of the more remote fisheries in the North East and can be found by taking the C-class road from the crossroads at the A68 and A692 junction at Castleside (Church Street) and following it up onto the moors. Continue on past Smiddy Shaw Reservoir, before bearing left approximately two miles after the turn-off for Waskerley village. The track is signed for Hawkburn.

West House Trout Lakes are situated beside the A66 between Stockton and Darlington. *Permits:* day tickets available on-site. *Species:* brown trout, rainbow trout, blue trout. *Website:* www.westhouse-troutlakes.co.uk

Witton Castle Trout Lakes are set in the grounds of Witton Castle, south of Witton-le-Wear, and controlled by Bishop Auckland and District Angling Club. *Permits:* day tickets are available on-site at the lodge. *Species:* rainbow trout. *Website:* www.bishopaucklandanddistrictanglingclub.co.uk/

STILLWATER COARSE FISHING VENUES IN NORTH YORKSHIRE

Broken Brea Fishery is located next to the B6271 between Brompton-on-Swale and the turn off for Easby, about two miles from Richmond. There are two lakes, one specimen and mixed coarse, and one for match-style fishing. *Permits:* day tickets available on-site. *Species:* carp, mixed coarse. *Contact:* telephone 01748 850107 or 07785 934371.

Woodlands Lakes is a highly regarded commercial day-ticket coarse fishery located immediately off the A61, between Thirsk and the A1, near to Carlton Minniot. There are thirteen well-stocked lakes, an on-site tackle shop and a café. *Species:* carp and mixed coarse. *Permits:* available on-site. *Contact:* telephone 01845 527099, 08731 824870, or email: info@woodlands-lakes. co.uk. *Website:* www.woodlandlakes-thirsk.co.uk/

STILLWATER GAME FISHING VENUES IN NORTH YORKSHIRE

Lockwood Beck Reservoir was formerly a part of the Northumbrian Water portfolio, but has for several years been leased by the Lockwood Beck Trout Fishery. In 2006 it was voted Britain's best trout fishery by readers of *Trout Fisherman* magazine and it is situated on the A171 between Guisborough and Whitby, approximately eight miles from Middlesbrough.

In common with a lot of the other 'independents', Lockwood Beck is fly fishing (with barbless hooks) only, all anglers must come equipped with a landing net and priest, and wading is restricted in certain areas. The fishery is stocked weekly with brown and rainbow trout ranging from 1½ lbs to double figures and permits vary in price from £640 to £210 per season for adults (prices dependent on the number of fish the angler expects to keep, with unlimited catch and release), £50 per season for children aged between twelve and sixteen, £24 for day tickets, £17.50 for evening permits and four-hour permits are £12. A day ticket for juniors (aged twelve to sixteen) is £10 and under-twelves can fish for free with an adult (all 2009 prices). There are eight rowing boats available for hire on this 60-acre fishery and the season runs from 4 April through to 31 October. Bookings and further information can be obtained from the fishery manager on 07973 779527 or email info@lockwoodfishery.co.uk

Scaling Dam is a popular multi-bait reservoir trout fishery controlled by Northumbrian Water and situated on the edge of the North York Moors. It is approximately 10 miles further east along the A171 from Lockwood Beck Reservoir. *Species:* rainbow trout, brown trout. *Permits:* available on-site. *Website:* www.nwl.co.uk/Exclusiveflyfisheries.aspx

Coarse and Game Fishing on Rivers in County Durham and North Yorkshire

COUNTY DURHAM

River Browney. The Malton & District Angling Club controls about 2 miles of the upper Browney (a tributary of the Wear) downstream from Lanchester. *Species:* brown trout (fly and worm). *Website:* www.communigate.co.uk/ne/maltonanglingclub/index.phtml

River Derwent. Forming the border between County Durham and Northumberland, the first 15 miles immediately below the dam of Derwent Reservoir (down as far as Lintsford) are under the control of the Derwent Angling Association, based in Shotley Bridge. The river contains brown trout and grayling and fishing is by fly only. Day tickets are available from the post office in Shotley Bridge or Frasers Angling, Coatsworth Road, Gateshead. Membership details can be found at: www.derwentangling.co.uk/. See also the entry for Northumberland and Tyne & Wear.

River Tees. Traditionally, the Tees formed the boundary between Durham and the North Riding of Yorkshire along its entire length, but following boundary changes in the 1970's the border moved south, transferring the upper Tees and its tributaries into a new and enlarged County Durham. Following the dissolution of Cleveland some twenty years later, the Tees does at least now constitute the boundary between North Yorkshire and Durham from Gainford to the sea and is home to a rapidly improving mixed coarse and game fishery throughout its length.

In the upper reaches, Upper Teesdsale Estates have approximately two miles of the north bank to the west of Middleton-in-Teesdale. Fishing is for brown trout, sea trout and salmon and permits are available from Raby Estate Office (near Staindrop on the A688 between West Auckland and Barnard Castle), High Force Waterfall Gift Shop and Raines Ironmongers in Middleton. Further details can be found by telephoning 01833 640209 or emailing teedaleestate@rabycastle.com

Further downstream, the Barnard Castle Angling Association has fishing on the Tees at Barnard Castle for brown trout, sea trout, salmon and grayling. Permits can be obtained from Wilkinson's Gun Shop, Castle Café or the Tourist Information Centre in the town.

Bishop Auckland & District Angling Club has a stretch of the Tees above and below Egglestone Bridge about two miles downstream of Barnard Castle. Target species include all those found at Barney, plus a variety of coarse species. Further details on Bishop Auckland's website: www.bishopaucklanda nddistrictanglingclub.co.uk/

Middlesbrough Angling Club has two stretches of the Tees in its middle and lower reaches, where the quality of the coarse fishing tends to be of a premium. At Over Dinsdale, near Darlington, and Stockton the key species are pike, barbel, chub, dace and roach. Permits for these stretches can be bought from Redcar Angling Centre, 159 High Street, Redcar, or Anglers Choice, 98 Cumberland Road, Middlesbrough.

See also Appendix Five for information on free fishing on the River Tees at four locations.

River Wear. There are some excellent stretches of mainly mixed game and coarse fishing under the control of several clubs on the River Wear. Durham City Angling Club controls fishing on three beats – Shincliffe/Maiden Castle, Cathederal/Prebends and Chester Moor. The stretch from Shincliffe to below Maiden Castle is an almost uninterrupted 3-mile continuum of first single- then double-bank fishing that can produce anything from Brown Trout to potential British record dace, double-figure sea trout to specimen barbel. chub are present throughout. The Cathederal/Prebends stretch consists of mainly deep, slow water beneath towering valley sides in Durham's iconic meander, yet it has an unlikely reputation for being one of the few places in Northern England where you can catch barbel in midwinter. Chester Moor is another fine beat for both coarse and game angling, found by taking the track that goes off the A167 Chester-le-Street bypass just behind Croxdale Autos. There are no day tickets for these waters, but Outside Visitor's Membership is offered at the reduced rate of £25 per season to anyone residing outside the post code areas CA, DH, DL, NE, SR, TD and TS. Information on all forms of membership of DCAC can found at: www.durhamanglers.co.uk/

Chester-le-Street Angling Club offers coarse and game fishing from Finchale Abbey, near Durham, almost all the way downstream to where the A1 motorway crosses the River Wear about a mile below Chester-le-Street. Day tickets are available for the C-L-S Park section and are issued by Chester-le-Street District Council from the Riverside Leisure Centre, located at the far side of Durham's Riverside Cricket Ground. CLSAC's website is www.chester-le-streetanglingclub.co.uk/

Bishop Auckland & District Angling Club offers 20 miles of fishing on the River Wear for salmon, sea trout, brown trout and grayling. The fishing beats of BAaDAC start in Weardale at Witton-le-Wear and stretch downstream to Croxdale, just outside Durham City. There is a lot of double-bank fishing, affording good access and the club welcomes members and visitors. *Website:* www.bishopaucklandanddistrictanglingclub.co.uk/

Ferryhill and District Angling Club is an established mixed fishing club with waters on the River Wear at Page Bank, Croxdale, and Chriton Avenue (Chester-le-Street), as well as a stretch of the Wear's tributary the River Gaunless. Information about membership can be found on the unofficial Ferryhill Members' website: www.ferryhillanglers.co.uk/

There are also several stretches of the Wear best suited to game fishing, mostly in the upper reaches in Weardale. The Upper Weardale Angling Association has about 6 miles of fishing for brown trout, sea trout and salmon from Cowshill in the upper dale down to Westgate. Permits can be obtained from the post office in St John's Chapel or the Blue Bell Inn.

Weardale Fly Fishers have 4 miles between Eastgate and Stanhope. Day tickets for that stretch can be bought from Stanhope Newsagents in Front Street, Stanhope.

See also Appendix Five for information on free fishing at two locations on the River Wear at Durham.

NORTH YORKSHIRE

River Esk. The Esk is officially North Yorkshire's only salmon and sea trout river and, with its only other natural inhabitant being the brown trout, this is truly a game fisherman's river. Egton Estates have one and a quarter miles of fishing with named pools available at Egton Bridge, but fishing is limited to threes rods per day. Booking in advance is therefore essential by phoning the estate office on 01974 895466.

River Leven. A tributary of the lower Tees, the Leven rises in the Cleveland Hills above Stokesley and enters the Tees on its Yorkshire bank. Middlesbrough Angling Club has a stretch of the river near to the confluence at Yarm, with brown trout, chub, roach and gudgeon the target species. Permits can be obtained from Redcar Angling Centre, 159 High Street, Redcar, or Anglers Choice, 98 Cumberland Road, Middlesbrough.

River Swale. The upper Swale is almost exclusively the preserve of the Richmond and District Angling Society. Their stretch of the river is effectively spit in two by Richmond Castle falls, the waters above the falls containing brown trout only and those from Mercury bridge, in Richmond, down, counting trout, grayling and numerous coarse species among their number. The deep pool beside the old station in Richmond, Easby Bend (about a mile downstream) and Great Langton, near Northallerton, are all locally renowned for their coarse (and in particular barbel) fishing potential. R&DAS membership is restricted to residents in the immediate Richmond area, but inexpensive day tickets can be obtained for all the club's waters from Gilsan Sports, 5 Market Place, Richmond, DL10 4HU, telephone 01748 822108.

There are two very good lower Swale day-ticket fisheries down the road that goes from the A168 near Asenby to Thornton Bridge and Helperby. The first, Cundall Lodge Farm, comprises about a mile of fishing in fairly slow, deep water that fishes well for barbel, chub, perch, pike and bream. Tickets (£5) are available at the farmhouse, which is by the main road just before the village of Cundall. Cars can be taken down and parked on the bank – turn in through the gate that is just before the house.

The second day-ticket fishery is just a couple of miles further downstream on the opposite bank and is controlled by the Helperby & Brafferton Angling Club. The water is next to the village of Fawdington. Day tickets (again £5) are available from the farmhouse, or alternatively from the village shop in Helperby (continue straight on for another mile after Thornton Bridge). Telephone 01347 821831.

River Tees. See entry for County Durham.

River Ure. The Ure, close to the small cathedral city of Ripon, is an excellent setting for a fine variety of river game and/or mixed coarse fishing, depending on your preference. The Ripon Piscatorial Association controls around 6 miles of double bank fishing up and downstream of the city, as well as fly-only fishing for brown trout in the Ure's small tributary, the River Laver. Membership details can be found on the association's website (www.ripon-piscatorial.co.uk), but £6 day tickets for the river can be bought from Bondgate Post Office, Ripon News (on North Street) and Ure Bank Caravan Site (all in Ripon) as well as Fish 'n' Things Tackle Shop, Horsefair, Boroughbridge.

Free Fishing Venues in the North East

There are only a few stretches of 'free fishing' available in the North East and all are parts of rivers – details below.

'Free fishing' means you can fish from a particular section of bank (or banks) without first having to get permission / a permit / a day ticket, provided you are in possession of a valid Environment Agency Rod Licence. Local Byelaws concerning seasons and rules that may apply at certain times of the year should always be consulted before fishing. Assume that the above is *not* the case on all other stretches of river in the region and *all* stillwaters!

*Note: The descriptions, 'right bank' or 'left bank' denote that particular bank when looking downstream.

FREE FISHING ON THE RIVER WANSBECK AT MORPETH

This section basically incorporates all accessible banks on the Wansbeck in the vicinity of Morpeth town centre. It includes the whole of Carlisle Park, High Stanners and the accessible sections of bank up and downstream of Telford Bridge (the main road to Newcastle).

Beginning at Lady's Walk, 100 yards upstream of the Skinnery footbridge, the right bank can be fished continuously downstream as far as the crest of the weir in Carlisle Park. The bank facing this is almost all private down as far as the Stepping Stones at High Stanners, except for a short section upstream of the stones that's accessible from the public footpath. The left bank downstream of the stepping stones can be fished, but access is tricky, as can the continuation of that bank below Oldgate Bridge down as far as the pleasure boat landing about 200 yards upstream of the weir (access less tricky). The park side of this section (right bank – 'The Prom') is open to fishing too, but the number of tourists in the area during high summer may make it impractical.

Adjacent to Chantry Museum, both banks between the footbridge and the main Telford Bridge can be fished, as can the right bank below the road bridge as far as the flood wall.

At Low Stanners, the left bank (only) is fishable from a point approximately 100 yards upstream of the metal footbridge until the sharp bend in the river several hundred yards downstream. This is the downstream limit of free water on the Wansbeck.

Species: brown trout, eels. *Season:* 22 March to 30 September inclusive. *Methods:* any, but no bait other than fly or worm before 1 June.

FREE FISHING ON THE RIVER WEAR AT DURHAM (1)

The downstream section of free fishing on the River Wear at Durham is one of the most famous (and popular) sections of river fishing in the region. It is known locally as both 'The Sands' and 'Feren's Park' and consists of a continuous stretch of the right bank, almost a mile in length, below Framwellgate Dam.

Access is by going down Claypath to the lights and turning right down the hill. The road gradually comes alongside the river and parking is available all the way along (charges apply in daytime). The lower boundary is adjacent to the Kepier Farm gate where the river begins to swing round to the left.

The reputation of the Feren's Park stretch for barbel fishing means it can get quite busy in the summer months. *Species:* barbel, chub, perch, bream (all to specimen sizes), brown trout, sea trout, dace, eels, tench (honest – I caught one here once!). *Seasons:* coarse species – 16 June to 14 March inclusive. Trout – 22 March to 30 September. *Methods:* any, but no coarse fishing tactics or baits, other than worm, are permitted between 15 March and 15 June inclusive.

FREE FISHING ON THE RIVER WEAR AT DURHAM (2)

Generally known as 'The Baths' stretch, this section of free fishing on the Wear runs downstream along the left bank from Durham Amateur Rowing Club, alongside the park, past the bandstand and Baths Bridge, to Elvet Bridge. It is generally regarded as coarse fishing territory, being of slower pace, with dace and roach the main species. *Access:* parking is available in Old Elvet. *Seasons and Methods:* as for The Sands / Ferens Park.

FREE FISHING ON THE RIVER TEES (1)

Barnard Castle, Co. Durham, 480 metres of the south bank downstream from the stone bridge to Thorngate footbridge. *Species:* mixed coarse and brown trout. *Seasons:* see River Wear. The taking of salmon is prohibited. The venue is just off the A67.

FREE FISHING ON THE RIVER TEES (2)

Gainford, County Durham. Fishing from church property only. The stretch is controlled by the Church Commissioners – strictly no fishing from Old Vicarage property. *Species:* mixed coarse, salmon and brown trout. *Seasons:* see River Wear. The venue is situated off the A67 at Gainford.

FREE FISHING ON THE RIVER TEES (3)

Hurworth Place, North Yorkshire (near Darlington). Fishing is from between the Skerne mouth and Croft Bridge. *Species:* mixed coarse, salmon and brown trout. *Seasons:* see River Wear. The venue is situated off the A167 at Croft.

FREE FISHING ON THE RIVER TEES (4)

Yarm, North Yorkshire. Fishing is on the Yorkshire bank (south) from about one mile upstream of the Yarm road bridge to a point 400 metres downstream of the same bridge. *Species:* mixed coarse. *Season:* 16 June to 15 March following – anglers fishing during the close season here have been successfully prosecuted by the Environment Agency in recent years. The venue is situated next to the A67 in Yarm.

References

BOOKS

Blakey, R., *Angling; Or, How to Angle and Where to Go*, Geo. Routledge, 1854.

Bradley, T., *Yorkshire Anglers Guide,* (2nd Edn), Leeds, 1894 (Republished Olena Books, 1979).

Charlton, C., and Byrne, P., *Jack Charlton: The Autobiography*, Corgi, 1997. Reprinted by permission of The Random House Group Ltd.

Hulme, M., *River Angling in Yorkshire*, 1997.

Johnson, F., *North East Angling Guide*, Northern Press, 1975.

Marshall, M. W., *Tyne Waters: A River and its Salmon*, Witherby, 1992.

Smith R., and Young, A., *Where to Fish, North East Stillwater Trout Fishing*, Sigma Leisure, 2001.

Various, *Angling in the Scottish Borders*, Borders Regional Council, 1979.

Young, P. (ed.), *Hooked on Scotland*, BBC/Mainstream Publishing, 1992.

MAGAZINES AND MAGAZINE ARTICLES

Clarke, Brian, 'Snaffle on Tees', *Waterlog*, Number 3, Spring 1997.

Clarkson, Geoff, 'Phoenix From The Flames', *Waterlog*, Number 45, April/May 2004.

Crawford, Lesley, 'Return of The Native', *Waterlog*, Number 38, February/March 2003.

Harwood, Keith, 'Bewick', *Waterlog*, Number 38, February/March 2003.

Mills, Derek, 'Dainty Dace', *Waterlog*, Number 50, February/March 2005.

Stirkazer, David, 'Lords And Ladies', *Waterlog*, Number 13, December/January 1998/9.

Tight Lines, The Newsletter of Durham City Angling Club, Issue 3, October 1993.

Tight Lines, The Newsletter of Durham City Angling Club, Issue 6, Summer 1995.

Tight Lines, The Newsletter of Durham City Angling Club, Issue 7, Autumn 1995.

Tight Lines, The Newsletter of Durham City Angling Club, Issue 9, Autumn 1997.

GOVERNMENT SCIENTIFIC REPORTS

River Coquet Salmon Action Plan Review, *APEM Aquatic Scientists (for The EA)*, February 2008.

River Tyne Salmon Action Plan Review, *APEM Aquatic Scientists (for The EA)*, February 2008.

River Wear Salmon Action Plan Review, *APEM Aquatic Scientists (for The EA)*, February 2008.

WEBSITES NOT CREDITED ELSEWHERE

www.waterlogmagazine.com
www.ne-fishing.co.uk
www.tyneriverstrust.org
www.fishpal.com
www.albagamefishing.com
www.dofreefishing.com
www.fishingarchives.com
www.flyforums.co.uk
www.barbel.co.uk

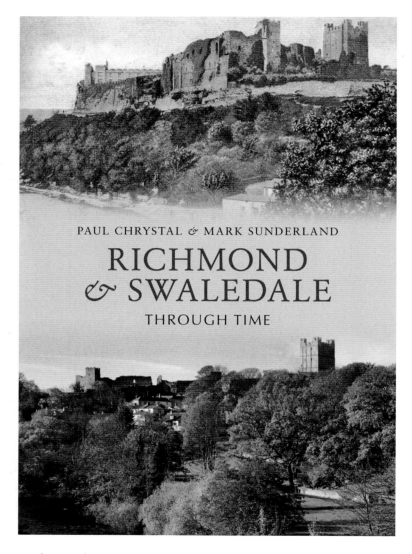

PAUL CHRYSTAL & MARK SUNDERLAND

RICHMOND & SWALEDALE

THROUGH TIME

Richmond & Swaledale Through Time
Paul Chrystal & Mark Sunderland

This fascinating selection of photographs traces some of the many ways in which Richmond and Swaledale has changed and developed over the last century.

978 1 84868 899 5
96 pages, full colour